INDIVIDUAL AND STRUCTURAL DETERMINANTS OF
ENVIRONMENTAL PRACTICE

Individual and Structural Determinants of Environmental Practice

Edited by

ANDERS BIEL
BENGT HANSSON
MONA MÅRTENSSON

Routledge
Taylor & Francis Group

LONDON AND NEW YORK

First published 2003 by Ashgate Publishing

Published 2017 by Routledge
2 Park Square, Milton Park, Abingdon, Oxfordshire OX14 4RN
711 Third Avenue, New York, NY 10017, USA

First issued in paperback 2017

Routledge is an imprint of the Taylor & Francis Group, an informa business

British Library Cataloguing in Publication Data
Individual and structural determinants of environmental
    practice. - (Ashgate studies in environmental policy and
    practice)
    1. Environmental psychology 2. Environmental protection -
    Citizen participation 3. Organizational behavior
    4. Sustainable development 5. Environmental policy
    I. Biel, Anders II. Hansson, Bengt III. Mårtensson, Mona
    337'.019

Library of Congress Cataloging-in-Publication Data
Individual and structural determinants of environmental practice / edited by Anders Biel,
    Bengt Hansson, Mona Mårtensson.
        p. cm. -- (Ashgate studies in environmental policy and practice)
    Includes bibliographical references and index.
    ISBN 0-7546-3217-2
        1. Environmentalism. 2. Behavior modification. I. Biel, Anders. II. Hansson, Bengt,
    1933- III. Mårtensson, Mona. IV. Series.

GE195.I53 2003
363.7--dc21

                                                                                2002043975

ISBN 13: 978-1-138-26419-9 (pbk)
ISBN 13: 978-0-7546-3217-7 (hbk)

Contents

PART II: COMMENTARY

Contributors

Anders Biel
Department of Psychology, Göteborg University
P.O. Box 500, SE-405 30 Göteborg, Sweden
e-mail: Anders.Biel@psy.gu.se

Carl Folke
Department of Systems Ecology, Stockholm University
SE-106 91 Stockholm, Sweden
e-mail: calle@system.ecology.su.se

Minna Gillberg
Department of Sociology of Law, Lund University
P.O. Box 114, SE-222 21 Lund, Sweden
e-mail: Minna.Gillberg@soc.lu.se

Bengt Hansson
Department of Philosophy, Lund University
Kungshuset, SE-222 22 Lund, Sweden
e-mail: Bengt.Hansson@fil.lu.se

Håkan Hydén
Department of Sociology of Law, Lund University
P.O. Box 114, SE-222 21 Lund, Sweden
e-mail: Hakan.Hyden@soc.lu.se

Arne Kaijser
Department of History of Science and Technology, Royal Institute of Technology
SE-100 44 Stockholm, Sweden
e-mail: arnek@kth.se

Mikael Klintman
Department of Sociology, Lund University
P.O. Box 114, SE-221 00 Lund, Sweden
e-mail: mikael.klintman@soc.lu.se

Anna-Lisa Lindén
Department of Sociology, Lund University
P.O. Box 114, SE-221 00 Lund, Sweden
e-mail: Anna-Lisa.Linden@soc.lu.se

Mona Mårtensson
Department of Sociology, Stockholm University
SE-106 91 Stockholm
e-mail: mmm@sociology.su.se

Ronny Pettersson
Department of Economic History, Stockholm University
SE-106 91 Stockholm, Sweden
e-mail: Ronny.Pettersson@ekohist.su.se

Rolf Wolff
School of Economics and Commercial Law, Göteborg University
P.O. Box 600, SE-405 30 Göteborg, Sweden
e-mail: Rolf.Wolff@handels.gu.se

Olof Zaring
Environmental System Analysis, Chalmers University of Technology
SE-412 96 Göteborg, Sweden
e-mail: olof.zaring@esa.chalmers.se

Preface

In the summer of 1995 nine funding bodies invited researchers in the social science and humanities in Sweden to take part in a new, large-scale and concerted research programme. This programme, called *Paths to Sustainable Development – Behaviour, Organization, Structure ('Ways Ahead')*, was launched with the aim of identifying and defining relevant research issues and paving the way for broader-based environmental research in the humanities and social sciences. One of the basic ideas was to establish environmental research more firmly in the disciplines where such work was needed, but where it was still relatively uncommon because environmental issues had simply not been recognized as relevant to the disciplines in question.

The contributors to this volume, with the exception of Carl Folke, have all been engaged in *Ways Ahead*. Their research has mainly been carried out in smaller groups including doctoral students at their own departments, but there has also been frequent common activities involving all groups. Part of what has been learnt is presented in this volume. The book brings forth the importance of attending to environmental practice at different levels in society. The common picture that emerges is that changing human behaviour in a more environmentally benign direction is not a matter of simple and sovereign policy that is effective for everyone, but a process, extended in time, with many potential obstacles. These insights pave way for multi-disciplinary research and realism in policy formation.

We are indebted to Professor Emeritus in Sociology Gösta Carlsson for valuable comments on an earlier version of this book.

March 2003

Anders Biel
Bengt Hansson
Mona Mårtensson

INTRODUCTION

Perspectives on Environmental Practice

Anders Biel

Many people are aware of such things as greenhouse gas emissions and depletion of the ozone layer. Some are also familiar with more specific details, for example that carbon dioxide and methane are gases that through emissions contribute to climate change and associated risks on earth. Not only may people have factual beliefs about problems impinging on earth. Many times they also have an apprehension about causes to these problems. Emissions from buildings, industries and transportation, leakage from cropland and landfills are sources that are widely recognized. Give it a little extra thought and an insight into one's personal contribution to existing problems may be realized.

Such familiarity indicates that natural science, in alliance with media, has had a vital impact on people's knowledge and everyday beliefs. Owing to natural science, problems are identified. Thanks to media, people are informed. Being enlightened citizens, they will then change their behaviour in a more environmentally benign direction. If not, blessed are the social scientists who will tell how things should be done.

So far, social scientists may not have come up to expectations. There are several accounts of this omission. People vary in their motives and goals. Some care more for personal consequences and their own well being while others have a stronger tendency to care for the collective good. Given that people vary in their value priorities, there is no single treatment that addresses all people. Moreover, value priorities are not displayed in a stable order across situations. Although thoughts about the environment may be alerted when throwing away refuse, this association is less likely to come to mind up in the air destined for holidays in the Caribbean. Nor is there a single treatment that addresses all mixed-motive situations. But even if people had the right minds, were prepared to prioritize environmental values and constantly thought about the environment, all situations or social structures do not leave room for displaying pro-environmental behaviour. Those living in a city with a well-developed underground system may, in order to promote pro-environmental behaviour, leave their private cars at home. Living in a city with no realistic travel mode options may force people to commute by car. Hence, large variations in motives and behaviour between individuals and between situations imply that in order to understand and predict human behaviour in environmental settings, a multi-disciplinary approach is called for.

A firm belief of the authors is that in order to reduce environmental problems human behaviour must change. However, as pointed out above behavioural changes have multiple determinants. Certain behaviours are guided by values and attitudes. Here, psychology can provide some answers about how to elicit values and influence attitudes that promote behavioural changes. Other actions are dictated by organizational norms and we can turn to studies in business administration to understand how management instruments develop and change. One should also be aware that environmental behaviour is shaped in the interplay between firms and customers, and between those in power and the governed. Sociology looks at how green identities are shaped in the relation between firms and their policies on the one hand and consumers that vary in identities on the other hand, while sociology of law addresses how the formation of legal rules and the application of law can promote sustainability among citizens. One must also keep in mind that many behaviours are shaped over time with little awareness on behalf of the actors about motives that guide behaviour. Thus, knowledge is required about contexts that form everyday practise. One such context is the physical and social world where we are brought up and where values are internalized. A sociological perspective can further our understanding of such matters. Another context is the physical structure that surround us in society, a structure that embraces a number of large technology systems for transportation, communication, and supply of services. Such systems have long term impact and a historical account could provide guidelines how to redirect them towards sustainability.

The Present Book

A distinguishing feature of this book is that it adopts a truly multi-disciplinary perspective on environmental problems and behavioural change, combining insight from research in sociology, psychology, business administration, sociology of law, history of technology, and philosophy. Thereby, this book brings forth the importance of attending to environmental behaviour at different levels in society: individual or household, organizational and societal. The book further accentuates that behaviours with negative or positive environmental effects are often performed without such consequences in mind. Moreover, this book clarifies why certain behaviours are rather easily changed while others are more firmly rooted in the individual make-up or in the social structure of society. Alas, there are no simple solutions to behavioural changes. Different actors are susceptible to different measures while at the same time a particular actor may be sensitive to different measures depending on which decision situation that arises. Furthermore, chapters in the present volume are guided by theoretical concepts within each subdiscipline. Such concepts are 'building blocks' that facilitate communication and integration between disciplines. Finally, and perhaps most importantly, the process character of behavioural change is emphasized throughout this volume. Changing human behaviour in a more environmentally benign direction is not a matter of a simple and sovereign policy that is effective for everyone, but a process, extended in time, with many potential obstacles.

The contributors to this volume, with the exception of Carl Folke, have all been engaged in the research programme *Paths to sustainable development – behavior, organizations, structure*, launched in 1996 by nine Swedish funding bodies. Their research has proceeded in smaller groups together with doctoral students at their own department, alternating with frequent common activities involving all groups. Part of what has been learnt is presented in the chapters to come.

Part 1
Environmental Practice: Individual, Organizational and Societal

A common understanding of human behaviour is that if only people are informed and knowledgeable, they will act in accordance with this new knowledge. In the chapter *Environmental Behaviour: Changing Habits in a Social Context*, Biel argues that many behaviours with environmental consequences are well-practised activities of the habitual kind. This is true for activities such as consumption behaviour and travel mode choice. Not only may people be less attentive to information targeted at the well-practised behaviour. Even if people are mindful of this information and form an intention to perform a new behaviour, this intention will be in 'conflict' with the old habit. As long as behaviours are performed in constant contexts, the intention to perform a well-practised activity may not be accessible to conscious awareness. Unless people are reminded about their recently formed intention to perform a new behaviour, the intention to perform the habitual behaviour may be triggered more or less automatically. In such potential conflicts, the habitual behaviour takes precedence. Whether people adopt a new behaviour or not is also influenced by what others do. If many around us practise a certain behaviour, this serves as a clue to proper behaviour. The mere fact that others continue to commute by car may silence one's conscience for doing the same.

In the next chapter, *Everyday Life Contexts and the Environment*, Mona

Mårtensson and Ronny Pettersson show how behaviours with environmental consequences are embedded in the practice and meaning of everyday life. They study the
importance for environmental practice of different everyday contexts, such as
people's way of defining resources and needs (cf. 'cultural theory'), gender order,
life course phase, life course experiences, physical infrastructure. The
authors demonstrate how combinations of these contexts vary between different
types of environmental practice. People's views of how pro-environmentalism
enters their everyday decisions are also analysed. The singling out of certain
areas or actions as long-term targets for environmental friendliness seems to be a
more frequent approach to caring for the environment than self-interest determined
by similar actions of significant others or internalized proenvironmental
principles.

There is a parallel in the findings about environmental behaviour by Biel on the
individual level and the findings by Mårtensson and Pettersson on the household
level. Both chapters emphasize that behaviours with environmental consequences
many times are embedded in everyday practice and not primarily dictated by an
environmental orientation or concern.

The following chapter, *The Formation of Green Identities – Consumers and
Providers*, addresses how green identities are formed among consumers and providers
of public services. Using electricity and waste as examples, Anna-Lisa Lindén and
Mikael Klintman argue that two important processes in greening provision and
consumption of goods and services in the public sector can be identified. One is the
identity formation by consumers and producers and the other process is materializing
public goods and services. Green identities can take on various forms and be driven by
different motives of individual and collective nature. The process of materialization
can be divided into three phases: make visible, acceptable, and 'doable'. Making
visible includes, among other things, to establish a green identity of the company.
Making acceptable comprises to reduce disincentives associated with higher prices
on green alternatives. Finally, making 'doable' can be exemplified by improved
conditions for recycling. Their conclusion is that both these processes are dependent
on the interplay between consumers and producers. Just as consumers can influence
producers to offer green alternatives, producers can actively shape people's identity
and product choice. To what extent these influences are forceful may vary between
services.

Turning to the private sector, in the chapter *Markets, Business and Sustainable
Repositioning*, Rolf Wolff and Olof Zaring present an analysis of sustainable
repositioning of industries. Here, they adopt an evolutionary perspective on this
process of change. As other 'organisms', companies react to environmental
pressure. During recent years they have been receptive to value changes in society
and the emphasis on the environment and sustainability. As a result, certain

companies have repositioned within the ecosystem they share with similar businesses. Based on their own studies, the authors argue that British Petroleum repositioned and selected a new niche, from oil producer to energy provider, within their ecosystem. In line with previous chapters, Wolff and Zaring accentuate the reciprocity between their study object and the surrounding society. To underscore this interplay they reason that due to greater pressure toward sustainability on the European market, as compared to the American, it is no coincidence that BP and Shell are up front in reshaping the agenda for the oil industry.

The final two chapters of part II emphasize the importance of structural factors in society in shaping long term patterns of environmental behaviour.

In the chapter *Legal and Governing Strategies – Towards a Law of Sustainable Development*, the sociologists of law Håkan Hydén and Minna Gillberg discuss the role of the law in sustainable development. Their point of departure is that the law should attend to effects of industrial activities in terms of consequences for the eco-system rather than worrying about the maximum costs that enterprises can bear to take precautionary measures. According to the authors a successful policy that meets acceptance includes, among other things, that the law should be reactive to tendencies in society. Let 'green' industries set the standard for legal measures, thereby forcing other companies to follow suit. In contrast to Hansson, see below, the authors argue that research should develop criteria for sustainable development. In this way scientific results can be directly fed into the political system and guide which standards the law should apply when evaluating, e.g., industrial enterprises.

In the chapter *Redirecting Infrasystems Towards Sustainability*, Arne Kaijser presents an historical account of the introduction and expansion of large technological systems, or infrasystems, taking gas lightning and transportation as examples. His historical exposé reveals that infrasystems are socio-technical systems in which the institutional frameworks and the system culture are as important as the technical components. Moreover, the development of infrasystems go through phases with different conditions; they interplay with each other, and they effect settlement patters. Hence, infrasystems have a vast and long-lasting impact on society and the environment. Based on historical lessons, Kaijser discusses how actions can be taken to redirect infrasystems in a more sustainable direction. Among other things, Kaijser reminds us that what is required to redirect infrasystems is not merely, and perhaps not primarily, a technical solution. Institutional frameworks and system cultures represent vital interests and die hard. Even though there are opportunities to redirect systems, it is a difficult task.

Part II
Commentary

Philosopher Bengt Hansson demonstrates the interplay between environmental research and politics in *A Dialogue Concerning the Usefulness of the Social Sciences*. This chapter is written in the form of a conversation between a Minister for the Environment and a social scientist. The Minister is worried that slow progress is made in changing people's environmental behaviour and at the same time somewhat irritated that social science research gives little guidance to bringing about a speedier process. Hansson argues that the primary task of social scientists is not to give definite recommendations to politicians. Rather, he shows how empirical findings can be useful in order to spot confounds and distinguish between factors with a persisting effect on the environment and factors that spuriously correlate with environmental consequences. Finally, Hansson reminds us that there are no problems that are simply environmental, but that societal problems carry with them consequences for a variety of sectors, the environment being one of them. An important point is that a wish to promote positive environmental consequences may be at odds with a desire to promote other goals. When our focus is on the environment, this is easily forgotten.

In the final chapter of the book, *Social-Ecological Resilience and Behavioural Responses*, the system ecologist Carl Folke presents a multidisciplinary view of sustainability and incorporates previous chapters into this framework. He emphasizes that socially determined habits of thought and action, reinforced by short-term successes, have lead to ecological and social vulnerability in the longer term. Folke discusses such vulnerability by means of the concept resilience. Resilience provides the capacity to absorb sudden change and to cope with uncertainty while maintaining desirable functions. The concept of resilience shifts perspective from the aspiration to control change in systems assumed to be stable, to sustain and enhance the capacity of social-ecological systems to cope with, and shape, change. The degree to which the social-ecological system can build and increase the capacity for learning, adaptation and responding in manner that does not constrain or end future opportunities is a central aspect of resilience.

PART I

ENVIRONMENTAL PRACTICE: INDIVIDUAL, ORGANIZATIONAL AND SOCIETAL

Environmental Behaviour:
Changing Habits in a Social Context

Anders Biel

Introduction

When I was a kid we learned to look to your right, look to your left and look to your right again before crossing the street. While learning, the new behaviour was part of an intentional system in which I consciously and deliberately formed a plan for performance. Over time, the behaviour came to operate autonomously. However, in 1967 Sweden changed to right-hand traffic and the rhyme had to be re-learned: look to your left, ... This spring and about to cross Tottenham Court Road in London I looked to my left and started crossing. This example illustrates that if the same kind of decision is made in a stable context habits may develop. It also elucidates that a functional behaviour in one time period may be less functional in another, and a behaviour that works in one setting may not be functional in a different setting.

Many behaviours with environmental consequences may be well-practised activities of the habitual kind. This is true for consumption behaviour (Dahlstrand and Biel, 1997; Grunert, 1996), recycling (Ouellette, 1996) and travel mode choice (Verplanken, Aarts and van Knippenberg, 1997). For many people these behaviours were established many years ago when environmental consequences was not an issue. Several years later they are asked to perform new kinds of behaviour such as to purchase eco-labelled products, to recycle into more and new fractions or to commute by public transport rather than by car. How could this come about?

A value that was not part of the original goal formulation should now influence the behaviour. Environmental considerations need to be taken into account when behavioural alternatives are evaluated. One problem with people who have established strong habits is that they are less likely to attend to information targeted at the well-practised behaviour (Verplanken and Aarts, 1999). Hence a new intention to behave in an environmentally friendly manner is not easily established. At the same time people should also get rid of the old behaviour. Even if a new intention is formed it might be in conflict with the old habit.

A model for behavioural change from an old to a new habit will be described below. Important factors in this process, besides attitudes, are environmental values and social norms. When environmental values are accentuated, the behaviour is placed within a social context. Not only personal but collective consequences become relevant. However, negative environmental effects result from the aggregated effects of many people's behaviour. The effects of one individual's behaviour are negligible. This question of personal efficacy suggests that it is important to establish standards for proper behaviour rather than focusing on personal contributions to the environmental problem at hand. Moral norms may function as a heuristic that guides behaviour. Moral norms prescribe that certain behaviours are inherently right or wrong, regardless of their personal or social consequences. They are social in their origin. When they become internalized, they have an autonomous influence over the individual's behaviour (Manstead, 2000). In this sense they function much like habits. Hence, new habits based on moral considerations may result in environmentally benign behaviour.

The next section provides a short review of recent research on automatic behaviour. This research underscores the functional aspects of such behaviour. However, changes in the social value structure may call for behavioural changes. The chapter proceeds with evidence relevant to factors that affect the process of changing habits. Placed in a social context, the importance of social norms for acting in a pro-environmental manner is recognized. How attitudes and norms may interact in shaping behaviour is then addressed. The chapter closes with some implications for environmental policy.

Automatic Behaviour

Reasoning

Within psychology the deliberate character of human decision making has been emphasized. This is true in attitude theory (e.g., theory of reasoned action) as well as in decision theories based on subjective expected utility. As Gigerenzer and Goldstein (1996) put it: '… the laws of human inference are the laws of probability and statistics' (p. 650). This standard view is taken to be both normative and descriptive of human inferences and decision making. Research on human heuristics and biases by Tversky and Kahneman (1974) seriously question the standard view as being a descriptive model of decision making. At the same time they maintain that the use of heuristics and biases in reasoning is a deviation from normative standards. Hence such reasoning is regarded as irrational. A different approach was taken by Herbert Simon (1982). He acknowledged that human reasoning capacities are limited but assumed that a limited capacity has ecological advantage. In a real-word environment with an enormous amount of information at each single point in time, and with an endless number of potential alternatives presenting themselves over time, humans need simpler procedures in order to get along. One such procedure, in contrast to optimizing, Simon called *satisficing*, implying that we choose an alternative that satisfies our aspiration level. Hence, we match the present situation with regularities of earlier experience.

This idea was fruitfully picked up by Gigerenzer and Goldstein (1996) who demonstrated the successful application of one satisficing principle, 'Take the Best', in human reasoning. This principle acknowledges that inferences are based on memory rather than from information presented in the situation. This algorithm can be applied in several steps. An example provided by Gigerenzer and Goldstein is to decide which of German cities has the larger population. In a first step a recognition principle could be applied. If one of two cities is recognized, choose the recognized city. If none or both are recognized, proceed and search for cue values. Here it is assumed that we have stored in memory cues that allow us to make inferences about cities. One such cue could be if a city is the capital or not. If we can discriminate based on this cue, choose the city that is known to be the capital. If not, retrieve a new cue and decide whether it discriminates or not. Hence, the algorithm is non-compensatory.

The 'Take the Best' strategy worked well in the task of picking out the largest city in Germany (Gigerenzer and Goldstein, 1996), but also in contexts such as selecting a share portfolio in a rising market (Gigerenzer, 2000). Intuitively, the recognition principle is a plausible one. When we go to the store, we many times choose among products on the basis of whether we recognize a product or not. We will emphasize this idea below when we introduce the concept of habits in purchase situations.

Mental Processes

Another line of research originates from cognitive psychology and the issue of conscious versus non-conscious processes in control of human behaviour. Here, Bargh (e.g., 1992, 1997) is a proponent of automatic processes that operate with a minimum of attention or awareness: '... much of everyday life – thinking, feeling, and doing – is automatic in that it is driven by current features of the environment (i.e., people, objects, behaviors of others, settings, roles, norms, etc.) as mediated by automatic cognitive processes of those features, without any mediation by conscious choice or reflection' (1997, p. 2). Such processes represent regularities of everyday experiences and they grow out of frequent and consistent experience. Automatic processes are many times proceeded by, or sometimes taken over by conscious processes (Wegner and Bargh, 1998). Conscious processes come in various disguise (ibid.). They can occur when people plan what to do, when they form an intention to behave, and when they need to monitor their behaviour. Planning involves both what to do among possible goals of action and how to achieve a given goal. Once such plans are executed repeatedly in stable contexts they become automated. However, now and then we need to attract consciousness to our actions. According to Wegner and Bargh this happens when an action is faulty and comes in two varieties; deliberate and event-driven monitoring. Deliberate monitoring occurs when an action is expected to be faulty or error-prone. One example is the construction of a difficult action sequence. Event-driven monitoring occurs when actions that are expected to go well somehow get off the track. About to pick up your diary from its ordinary place on your desk you may realize that it is gone.

Habits

The idea that automatic processes can be more efficient than slower, controlled, processes was emphasized by William James as early as 1890, although James used the notion of habit rather than satisficing algorithm or automaticity. The more of habits the better was his motto. The mind should be set free for proper work of higher powers. And unfortunate were those who did not form habits: 'There is no more miserable human being than one in whom nothing is habitual but indecision, and for whom the lighting of every cigar, the drinking of every cup, ..., are subjects of express volitional deliberation' (James, 1948, p. 145). According to James, habitual behaviour is learned in a social context. Furthermore, the function of habits is to use our mental capacities prudently and to simplify our actions. This view of habits is well in accordance with modern psychology on automatic mental processing and habitual behaviour.

Despite an early take-off, the study of habits has been largely ignored in social psychology. With the cognitive revolution and an interest in cognitive processes, expectancy-value models such as the theory of reasoned action (Ajzen and Fishbein, 1980) and the theory of planned behaviour (Ajzen, 1985, 1991) dominated the field. Behaviour is seen as intentional and attitudes shape our intentions. Nevertheless, some researchers took an interest in habits. Triandis (1977, 1980) proposed that attitudes and intentions may cause action when the behaviour is new. When the same behaviour is repeated habit grows stronger and the effect of intention loses its significance. This has later been demonstrated (Verplanken et al., 1998). Other studies have shown that past behaviour predicts future behaviour over and above factors included in the theory of reasoned action and the theory of planned behaviour (e.g., Bagozzi, 1981; Bentler and Speckart, 1979; Mittal, 1988).

Recent reviews of habits versus intentional behaviour (Ouellette and Wood, 1998; Verplanken and Aarts, 1999) also speak to the importance of habits in predicting future behaviour.

Research on automatic reasoning and action shows that such behaviour is functional in our every-day lives. Habitual behaviour is an automatic response to cues in the environment, it proceeds with little awareness and is goal-directed. Since habits are functional, there is little internal motivation to change behaviour. Furthermore, external attempts to change habits may have little success. When habits are strong, people attend less to contextual information (Verplanken, Aarts and van Knippenberg, 1997). They are also less prone to consider alternative courses of action (ibid.). Habitual change is also a process where newly formed intentions have to 'compete' with the old habit. Even highly motivated people may find it difficult to break habits. This is not to say that people never break their habits. Several conditions may have to be fulfilled though for a positive transition to new and more environmentally benign habits. This process will be described below with examples from research on consumer behaviour.

Breaking Habits

The purchase of many everyday products has an habitual character. It is performed in a stable context, often executed with high frequency and without much reflection. A familiar brand label or product look may serve as a cue initiating an automatic response or habit. During recent years consumers have been asked to show environmental concern. The automatic process should compete with a new behavioural intention: to make an environmentally benign choice. A frequent and successful implementation of this new intention will, it is hoped, result in a new habit that replaces the old one.

This process of habitual change has been investigated in several survey studies (Biel and Dahlstrand, 1999; Dahlstrand and Biel, 1997; Grankvist and Biel, 2000a, b). A model that traces this process of changing an old habit into a new one was initially presented in terms of seven steps (Dahlstrand and Biel, 1997). The broad outline of this model can be described in three phases. General factors such as environmental values and a sense of personal responsibility for contributing to environmental damages are influential in an early phase of transition. This environmental awareness is paralleled with a positive general attitude toward ecological behaviour and an impetus to attend to your present behaviour. In a second phase people take a new behaviour into consideration and examine alternative ways to perform this new behaviour. Such considerations affect specific attitudes and beliefs. Specific beliefs are beliefs about performing a behaviour with regard to the object; e.g., to purchase an ecological product. As an example, you may have to pay more for an ecological than an 'ordinary' product or it may be more difficult to find in your shop. Thus, specific beliefs are here assumed to influence the propensity to try out and to buy a particular product. Once the new behaviour is tested it is also assumed to be evaluated, the third phase. Since many values may affect this evaluation, it is considered to be of special importance that environmental values guide further purchase behaviour.

Dahlstrand and Biel (1997) found that environmental values were important in a first phase of transition. By contrasting a group of individuals who never bought eco-labelled washing and washing-up detergents with a group that had given it serious thoughts, they showed that individuals in the second group rated environmental values as more important than individuals in the first group. In a follow-up study with a wider range of products, the importance of environmental values in an initiating phase of changing habits was confirmed (Grankvist and Biel, 2000a).

The second phase of behavioural change implies that people consider alternative behaviours. They are also more attentive to specific information about these new behaviours. In our studies of purchase of eco-labelled products (Dahlstrand and Biel, 1997; Grankvist and Biel, 2000a, b) specific beliefs about eco-labelled products affected whether people would form a more positive attitude towards testing such products. For example, those respondents who believed that they did not have to pay more for eco-labelled products than ordinary products, or that labelled products were better for their personal health, were more likely to try them out. In the third phase the new behaviour is practised and evaluated. Evidently, a more positive evaluation will have a more positive effect on the future likelihood of practising the new behaviour and vice versa. Importance attached to environmental values as well as a general positive attitude towards eco-labelled products was shown to increase the likelihood that respondents continued to purchase organic food products rather than non-organic alternatives (Grankvist and Biel, 2000a).

Not all people proceed on the route towards buying more environmentally friendly products. Some never take the first step. This is true in particular for people with weak environmental values. Some start off but relapse into their old habits. The panel study by Grankvist and Biel (2000b) gave some hints as to why this is the case. These results are much the mirror image of progress. Those among the respondents that at the start of the study purchased eco-labelled products, regularly or now and then, but one year later only did it now and then or never, attached less value to the environment than did participants that remained at the same purchase level or increased their share of eco-labelled products. The former group also had a less positive attitude towards such products.

To summarize, some consumers never associate environmental values with attitudes toward purchase of everyday products. Other criteria will determine their choice. Many additional consumers have only weak associations between environmental values and attitudes and behaviour. They will not actively search for ecological products. However, a reminder in the purchase situation may activate a link between value and attitude. A prompt such as an eco-label provides a cue to a promising alternative. Finally, some consumers have internalized environmental values and look out for ecological products.

A Social Context

So far the process of behavioural change has been described as an individual endeavour. However, most behaviours take place within a social context. People may come to an understanding that their actions have negative consequences for the environment. At the same time, many people contribute to the aggregated negative effects. This is a situation that is true for many acts with environmental consequences. On the one hand many individuals must change their behaviour. On the other hand there is little incentive for a particular individual to change his or her behaviour in order to promote the collective good. Rather, individual incentives could be more forceful. Commuting by bus versus a quicker and more comfortable journey by car could speak in favour of the latter alternative. To buy an eco-labelled product often comprises an extra cost compared to the choice of an 'ordinary' product. At the same time, if most people follow their individual incentives, all may be worse off than if they had co-operated in the first instance. Such conflicts between private interests and the interest of the collective at large has the structure of a social dilemma (Dawes, 1980).

What then could foster co-operation in social dilemmas? Among factors that have been shown to yield positive effects are a larger pay-off, communication among group members, reduced social and environmental uncertainty, self-efficacy, and social norms (Komorita and Park, 1994; Ostrom, 1998; van Lange et al., 1992). These findings are primarily based on experiments in the laboratory. However, many of these factors are absent in social environmental dilemmas (Biel, 2000). There is little room for increased self-efficacy. No matter how much one individual tries, the effects on the environment will be negligible. To make things worse, many times there is uncertainty about the aggregated effects on the environment (environmental uncertainty) as well as a lack of feedback about the negative consequences of individual behaviour. Taken together, values, attitudes, and beliefs about consequences may not suffice for a behavioural change. Among other factors, rules or norms about proper behaviour could be a vital supplement.

Social Norms

Earlier research points to the importance of prescriptive norms for cooperation in social dilemmas, how one ought to act in a specific situation. In particular, the norm of reciprocity has drawn a wide attention. Kerr (1995) suggested the importance of reciprocity in experimental studies of social dilemmas. If others are expected to cooperate, and hence make a sacrifice for the common good, the norm requires reciprocation. Ostrom's review of an abundant research of experimental social dilemma studies (1998), as well as her own studies of real-life, local, common pool resource dilemmas (1990), show that the norm of reciprocity is powerful. Further support comes from computer tournaments in a Prisoner's Dilemma game where the tit-for-tat strategy outperformed other strategies (Axelrod, 1984).

Two conditions seem to contribute to the success of the reciprocity norm, either in isolation or together. One is that people have face-to-face communication, the other that one can monitor what other people do (Biel, 2000; Ostrom, 1998). In many large-scale dilemmas, where people act under high anonymity, low group solidarity, little personal communication, and with low personal efficacy, these conditions are hardly to be found (Kerr, 1995). This absence of vital conditions, together with results from earlier research (Biel, von Borgstede and Dahlstrand, 1999; Nielsen and Hopper, 1991; Schwartz, 1977; Stern et al., 1999; von Borgstede, Biel and Dahlstrand, 1999), led Biel (2000) to suggest that prescriptive norms about what one ought to do as a citizen, a social duty, rather than norms governing interpersonal interactions, such as reciprocity, promote cooperation in large-scale dilemmas with environmental consequences.

In his norm-activation theory of helping behaviour Schwartz (1977) recognized that altruistic concerns about other people may activate feelings of a moral obligation to help. Schwartz's theory was later applied to recycling behaviour (Nielsen and Hopper, 1991) where a moral obligation to recycle promoted pro-environmental behaviour. Stern et al. (1999) extended Schwartz theory to include not only people but also other valued objects such as tropical forests and other species. Their research indicate that moral norm is an important factor when it comes to support environmental movement. Further evidence comes from studies of behaviour in every-day environmental dilemmas (Biel et al., 1999; von Borgstede et al., 1999). Vignettes of environmental dilemmas were presented to the respondents. For example, households could be asked to reduce their electricity consumption during a shortage. A set of questions were asked for each situation. One question measured their likelihood of changing their present behaviour. Other questions concerned the need for a reduction of electricity use in society, the importance and seriousness of problems created by an extensive use of electricity, their own responsibility for the potential situational problem, the significance of various consequences for themselves and for others from saving or not saving electricity, and perceived moral norm strength that they should reduce their electricity consumption. In both studies there was a high positive correlation between perceived norm strength and respondents' willingness to behave in a more pro-environmental manner. Hence, the stronger a moral norm is perceived in a large-scale environmental dilemma, the more likely people are to cooperate for the common good.

It should be noticed that these findings were based on correlations. Respondents could have adjusted their estimates of norm strength to their own willingness to cooperate. To gain better control of the causal order, moral norm strength was manipulated in an experiment (von Borgstede et al., 1999, Experiment 1). Here, the number of the population that supported a moral norm varied between 20 per cent and 80 per cent in four different situations. When respondents believed that the support was strong, more were prepared to cooperate than when they believed that a minority of the population subscribed to the norm.

One possibility is that respondents had no personal opinion about what was the right thing to do in these situations. Once they learned what others thought, this was all that mattered. However, separate measures showed that they had an opinion of their own. Furthermore, this personal norm strength did not vary with social support. While people may have their own view about the right thing to do, they are also influenced by what others think one ought to do. We believe that the general level of cooperation is influenced by the reason people have that others will defect. An injunctive norm can serve as a proxy for what kind of action other people take. If it is perceived as weak there is less reason to believe that others will cooperate. If it is perceived as strong others are expected to act accordingly. In short, social uncertainty is reduced through knowledge about the support of moral norms in society.

In their model of habitual change Dahlstrand and Biel (1997) recognized that social norms could have an impact on the process. Two kinds of norms were suggested to have an effect; prescriptive and descriptive social norms. While a prescriptive norm refers to what people should do in a particular situation, a descriptive norm describes what is typical or normal behaviour in a specific situation. Social uncertainty could also be reduced by information about what other people actually do (von Borgstede et al., 1999, Experiment 2). This could be of special importance in the phase where people start to perform a new behaviour. If a person believes that many others perform the new behaviour, this may affect the probability that the person him- or herself behaves in a similar fashion (Kerr, 1995; Klandermans, 1992). Not only can one trust that others cooperate, but also that in the aggregate the new behaviour might have a positive effect on the environment.

Individual Consequences in a Social Context

The construct of moral norm has been attended to in models of attitude-behaviour relation. As is clear from the review by Manstead (2000), moral norm serves a parallel function to attitudes and can influence behavioural intention independent of other constructs included in theories such as the theory of reasoned action or the theory of planned behaviour. Here it is suggested that attitudes and norms sometimes can contradict each other in shaping environmental behaviour.

Frey and Oberholzer-Gee (1997) analysed the reaction to a nuclear waste repository among citizens living in a prospective host community. In spite of the fact that a nuclear waste repository is seen as a negative facility by residents, more than half of the respondents (51 per cent) agreed to have the repository placed in their own community. In the next phase of the study Frey and Oberholzer-Gee introduced a compensation package. They explained that the Swiss parliament had decided to compensate all residents in the host community. Respondents were offered a different sum per household and year for acceptance (CHF 2,500, 5,000 and 7,500, respectively). Although such a compensation package is supposed to increase the acceptance of a noxious facility, the authors made a different prediction. The results first. When money was offered the acceptance rate dropped to around 25 per cent. The amount of compensation had no effect on the acceptance rate. In line with crowding theory (for a summary; see Frey, 1997) intrinsic motivation to accept the facility was reduced or crowded out by the monetary compensation. Verbal comments by the respondents to their decision substantiated that different motives affected the decision with and without compensation. Without compensation, the degree of opposition toward nuclear power and acceptance of the current political citing procedure affected the acceptance to host the waste repository. When compensation was introduced, the effect of these factors was absent and only individual consequences (personal risk and economic impacts) determined the decision. Rather than being a social and societal problem where a sense of civic duty guided the decision, individual incentives came in focus.

A second example with a similar message comes from a study by Tenbrunsel and Messick (1999). Two groups of subjects role-played managers that were asked to allocate part of their budget in running scrubbers that would reduce emissions. If most managers did so, the goal of the company would be reached. In one of the groups a weak sanctioning system was introduced. If they did not comply with the company's policy, there was a small risk that they would be met with sanctioning costs. In the group without sanctions around 75 per cent of the managers co-operated while less than 50 per cent did so in the sanctioning group. The explanation provided by the authors is that without sanctions, the decision is seen as an ethical one. The right action is to avoid cheating. However, when sanctions are introduced this becomes a business decision and because the sanction costs are so low, the scrubbers are cheaper to turn-off than run.

These two studies show a situation where individual and social consequences are at odds. Economic incentives speak to individual consequences while moral norms address social values. When the focus is on social and moral aspects of the decision, respondents show a positive attitude toward the pro-social choice. When economic incentives are introduced, the balance shifts in favour of recognizing individual consequences. As a result, the attitude toward the pro-social choice is less positive.

In other situations it could be difficult for moral and social consequences to even have an impact on behaviour. It is not sufficient that present behaviour may have serious negative effects on the environment. The Swedish public for example experience air pollution from car traffic as a serious problem today which will persist over the next ten years. The same is true for the use of pesticides in farming (Biel, von Borgstede and Dahlstrand, 1999). However, while people perceive a moral obligation to buy organic food in order to reduce the use of pesticides, they fail to reduce their car travel. People must also recognize a need for behavioural changes in society and the importance of behavioural changes among various groups in society (ibid.). Furthermore, they have to sense a personal responsibility for the problems at hand (Schwartz, 1977). Why then are these conditions fulfilled for buying organic products but unfulfilled when reducing car travel?

A study of private car use (Bennulf et al., 1998) suggests that a change of present behaviour may be seen as unfeasible. The cost in terms of individual consequences is too high. Several measures were included in this study such as general environmental attitudes and values, attitudes towards traffic and towards environmental consequences from traffic, habitual car use, and every-day need of their car. The set of attitude variables did not explain any of the variance in car use while habit and every-day need were strong predictors. Attitudes and values seemed to be too weak to support the evolvement of a prescriptive norm to avoid car use that could bolster the influence of positive individual consequences from present car use. A recent study of behavioural changes in the office (von Borgstede and Biel, 2000) confirms this interpretation. In this study behaviours were divided into those that were easier to change and those that were more difficult to change. A moral norm to change behaviour was perceived as stronger in the former category, e.g., that one should switch off the computer screen while not in use, than in the latter, e.g., that one should travel by train rather than by air to a long distance meeting.

Conclusions and Implications for Environmental Policy

'If only people are informed, if only people have the right knowledge...' The 'if' implies that then they will adopt their behaviour accordingly. But many people do know about organic food and even something about differences in production methods as compared to regular food products. Nevertheless, a minority purchases organic food products. People also know about adverse consequences from green-house gases and about the contribution of car traffic to such emissions. Still, there is no reduction in transport behaviour in sight. Does this mean that information is useless? A safe answer is no, but it needs to be qualified.

One thing that complicates matter is that many behaviours with environmental consequences are habitual in nature. They are functional and there is little inner motivation to change behaviour. Studies have shown that when this is the case, people attend less to relevant information than if behaviour is guided by conscious processes. Take as a related example somebody who is thinking of moving to a new flat. This person is probably much more likely to look for advertisments about flats in the newspaper than someone who is happy with her present residence. Information becomes relevant when we need it and when it is of value to us. One could argue that since many people value the environment, environmental information is also of value. Hence, people would find it valuable to know about new and more environmentally benign production methods in farming or about greenhouse gas emissions and their negative effects on global climate. But do people believe that they need such information in the sense that it says something relevant about their present behaviour? Unless people understand that their present behaviour contributes to negative environmental consequences, they may not see the relationship between their choice of 'ordinary' food products and environmental effects or the relationship between car use and emissions. Specific information about such relationships is more important than general information about detrimental effects on the environment without specifying the causes.

But even if people are aware that a behavioural change is called for, they may still need information about how to proceed. In the case of purchase, which products do I look for, where do I find them, and what do they look like? Since there is nothing in the product itself that signals organic food or 'good environmental choice' to the consumer, a special label could help the consumer to discriminate between products. Provided that there is an agreed upon eco-labelling system, such information is easily conveyed. Information about alternative travel modes could also be presented in an intelligible way. Maps incorporating bike lanes and bus routes, complemented with time tables, can be distributed to the households. Notice though that in order for procedural information to have an impact on behavioural change, people must first accept that a change on their behalf is called for.

However, this is where acts such as consumer and transport behaviour can be distinguished. While information may contribute to a voluntary change of behaviour with regard to consumption, coercion seems unavoidable in the latter case. As was pointed out above, environmental acts can differ with regard to how much people have to adjust in order to promote the collective good. To choose a new product requires little amendment. Once people recognize an eco-label, the labelled product could become a promising option. However, a direct comparison with an equivalent, non-labelled, product may speak in favour of the latter. Hence, to choose an eco-labelled alternative may be seen as a minor self-sacrifice. This is where information about the behaviour of others could be salient. Unless one can trust other people to co-operate, a self-sacrificing act will have little or no effect. Information in media, and in the shops, about the behaviour of the collective at large, or what a majority of the collective regards as prudent behaviour, can influence personal behaviour in a pro-environmental direction.

When it comes to parking your car and use other transport means, the behavioural change is more overturning and environmental concern is not easily manifested in pro-environmental behaviour. In many instances, alternative transport modes may not exist. And even when they do, a trip may involve a sequence of errands rather than just going from point A to point B and back again. Procedural information may describe alternative travel modes in a clear way. Still, to change from car to e.g. bus involves much more of planning than to switch from buying non-organic products to buying organic ones. Furthermore, changes in personal consequences are much more evident in the former case than in the latter. A comparison may speak in favour of the car alternative. Finally, feedback about others' behaviour is more manifest on the road than in the shop. Since roads are crowded, a norm against driving is hard to imagine; rather the opposite.

Just as politicians realize that road transport cannot be banned since other goals and values are in conflict with environmental ditto, they must understand that the public have other goals in mind when they act than protecting the environment. To go by car could be anchored on gratification for oneself in terms of comfort and time saving. Food choice is many times associated with taste and pleasure and with health and security. A change in value priorities is a time-consuming process.

References

Ajzen, I., (1991), 'The theory of planned behavior', *Organizational Behavior and Human Decision Processes, 50*, 179–211.

Ajzen, I., and Fishbein, M., (1980), *Understanding attitudes and predicting social behavior*, Englewood Cliffs, NJ: Prentice-Hall.

Axelrod, R., (1984), *The evolution of cooperation*, New York: Basic Books.

Bargh, J. A., (1992), 'The ecology of automaticity: Toward establishing the conditions needed to produce automatic processing effects', *American Journal of Psychology, 105*, 181–199.

Bargh, J. A., (1997), 'The automaticity of everyday life', in R. S. Wyer (ed.), *The automaticity of everyday life: Advances in social cognition*, Vol. 10, pp. 1–61, Mahwah, N.J.: Lawrence Erlbaum.

Bennulf, M., Fransson, N., Polk, M., and Biel, A., (1998), *Automobility and environment: Attitude and attitude formation* (KFB-report 4, in Swedish), Stockholm: The Swedish Transport and Communications Research Board.

Bentler, P. M., and Speckart, G., (1979), 'Models of attitude-behavior relations', *Psychological Review, 86*, 452–464.

Biel, A., (2000), 'Factors promoting cooperation in the laboratory, in common pool resource dilemmas, and in large-scale dilemmas: similarities and differences', in M. Van Vugt, M. Snyder, T. R. Tyler, and A. Biel (eds), *Cooperation in modern society. Promoting the welfare of communities, states and organizations*, London: Routledge.

Biel, A., and Dahlstrand, U., (1999), *Habits and the establishment of ecological purchase behaviour* (Manuscript).

Biel, A., von Borgstede, C., and Dahlstrand, U., (1999), 'Norm perception and cooperation in large scale social dilemmas', in M. Foddy, M. Smithson, S. Schneider, and M. Hogg (eds), *Resolving social dilemmas: Dynamic, structural, and intergroup aspects*, Philadelphia: Psychology Press.

Dahlstrand, U., and Biel, A., (1997), 'Pro-environmental habits: Propensity levels in behavioral change', *Journal of Applied Social Psychology, 27*, 588–601.

Dawes, R. M., (1980), 'Social dilemmas', *Annual Review of Psychology, 31*, 169–193.

Frey, B. S., (1997), *Not just for the money. An economic theory of personal motivation*, Cheltenham: Edward Elgar.

Frey, B. S., and Oberholzer-Gee, F., (1997), 'The cost of price incentives: An empirical analysis of motivation crowding-out', *American Economic Review, 87*, 746–755.

Gigerenzer, G., (2000), *Adaptive thinking: Rationality in the real world*, New York: Cambridge University Press.

Gigerenzer, G., and Goldstein, D. G., (1996), 'Reasoning the fast and frugal way: Models of bounded rationality', *Psychological Review, 103*, 650–669.

Grankvist, G., and Biel, A., (2000a), *The importance of beliefs, purchase criteria and habits for the choice of environmentally friendly food products* (Manuscript).

Grankvist, G., and Biel, A., (2000b), *Unfreezing habits and activating purchase criteria* (Manuscript).

Grunert, K. G., (1996), 'Automatic and strategic processes in advertising effects', *Journal of Marketing, 60*, 88–101.

Hopper, J. R., and Nielsen, J. M., (1991), 'Recycling as altruistic behavior: Normative and behavioral strategies to expand participation in a community recycling program', *Environment and Behavior, 23*, 195–220.

Kerr, N. L., (1995), 'Norms in social dilemmas', in D. Schroeder (ed.), *Social dilemmas: Social psychological perspectives*, pp. 31–47, New York: Pergamon Press.

Klandermans, B., (1992), 'Persuasive communication: Measures to overcome real-life social dilemmas', in W. Liebrand, D. M. Messick and H. Wilke (eds), *Social dilemmas: Theoretical issues and Research Findings*, pp. 307–318, Oxford: Pergamon.

Komorita, S., and Parks, C. D., (1994), *Social dilemmas*, Madison: Brown and Benchmark.

Manstead, A., (2000), 'The role of moral norm in the attitude-behavior relation', in D. J. Terry and M. A. Hogg (eds), *Attitudes, behavior, and social context*, pp. 11–30, Mahwah, N.J.: Lawrence Erlbaum.

Ostrom, E., (1990), *Governing the commons. The evolution of institutions for collective action*, New York: Cambridge University Press.

Ostrom, E., (1998), 'A behavioral approach to the rational choice theory of collective action', *American Political Science Review, 92*, 1–22.

Ouellette, J. A., and Wood, W., (1998), 'Habit and intention in everyday life: The multiple processes by which past behavior predicts future behavior', *Psychological Bulletin, 124*, 54–74.

Schwartz, S. H., (1977), 'Normative influences on altruism', in L. Berkowitz (ed.), *Advances in Experimental Social Psychology, 10*, 221–279.

Tenbrunsel, A. E., and Messick, D. M., (1999), 'Sanctioning systems, decision frames, and cooperation', *Administrative Science Quarterly, 44*, 684–707.

Triandis, H. C., (1980), 'Values, attitudes, and interpersonal behavior', in H. E. Howe, Jr., and M. M. Page (eds), *Nebraska symposium on motivation, 1979*, Lincoln, NE: University of Nebraska Press.

Tversky, A., and Kahneman, D., (1974), 'Judgment under uncertainty: Heuristics and biases', *Science, 185*, 1124–1131.

Van Lange, P. A. M., Liebrand, W. B. G., and Messick, D. M., and Wilke, H. A. M., (1992), 'Introduction and literature review', in W. Liebrand, D. M. Messick and H. Wilke (eds), *Social dilemmas: Theoretical issues and Research Findings*, pp. 3–28, Oxford: Pergamon.

Verplanken, B., and Aarts, H., (1999), 'Habit, attitude, and planned behavior: Is habit an empty construct or an interesting case of goal-directed automaticity?', *European Review of Social Psychology, 10*, 101–133.

Verplanken, B., Aarts, H., and van Knippenberg, A., (1997), 'Habit, information acquisition, and the process of making travel mode choices', *European Journal of Social Psychology, 27*, 539–560.

Von Borgstede, C., Dahlstrand, U., and Biel, A., (1999), 'From ought to is: Moral norms in everyday social dilemmas', *Göteborg Psychological Reports, 29*, No. 5.

Wegner, D. M., and Bargh, J. A., (1998), 'Control and automaticity in social life', in D. T. Gilbert, S. T. Fiske and G. Lindzey (eds), *The handbook of social psychology*, Vol. 1, pp. 446–496, Boston: McGraw-Hill.

Everyday Life Contexts
and the Environment

Mona Mårtensson and Ronny Pettersson

Introduction

The question of what can incite people to act in the interests of the environment in their everyday practice is central to research into methods of achieving ecological sustainability. One important consideration is how respect for the environment is incorporated, or embedded, into the practice and meaning of everyday life. On a meanings level, embeddedness is about the adapting of environmental orientation to other interests, reasons and motives.

Moreover, for many it is not only personal motives that come into play in their everyday practice, but personal motives modified to the values and wishes of other members of their immediate environment, primarily those in the same household. The balance between different motives in the actions and choices of the household are often a product of such compromises. This makes it especially difficult to unravel the relationship between the different motives behind people's everyday practice.

The purpose of this chapter is to try to understand environmental orientation as embedded in different kinds of context. We will be looking at everyday life contexts at a given point in time as well as at processes over an individual's life course. We will also be looking at contexts detectable in people's declared actions as well as at those based on their descriptions of how they take the environment into consideration in their everyday practice. We will firstly be discussing the part the environment plays in the practice and meaning of everyday life as suggested by an extensive survey of households in Stockholm, the capital of Sweden. The points of departure will be a broad theory of ways of life and two explanatory models that focus on certain dimensions of the everyday context, namely life course phase and gender order. Following this, we will look at the context that comprises a person's previous experiences. Finally we will be addressing another kind of meaning context, namely people's own reports of how they incorporate proenvironmentalism into their everyday decision-making processes. The discussion of these two latter contexts is based on a study carried out on a smaller number of Swedish households.

Some Thoughts on Environmental Oriented Everyday Life

Socio-Economic Characteristics, Age and Gender

In previous studies, an individual's environmental commitment and proenvironmental activities have often been put in relation to variables such as socio-economic characteristics, age and gender.[1]

According to one once widely-held idea, well-educated people and those with a relatively good *socio-economic status* show more of an interest in proenvironmental alternatives than others (Bennulf, 1994, p. 117; Lindén, 1996, p. 87; Hallin, 1994). One common conception has been that interest in the environment increases with income, with those who are less well-off thinking primarily of improving their material well-being. Environmental concern is seen as a luxury. There is nothing, however, to suggest that environmental concern increases with socio-economic resources (Brand, 1997). A related hypothesis in previous research has been that proenvironmental attitudes and activities are mainly the reserve of the well-educated. This claim is based on the idea that the well-educated, by virtue of their knowledge and being accustomed to complex abstract reasoning, find it easier than others to take into consideration the long-term, deferred effects of their own actions. This is also a widely held conviction with the public at large, expressed in statements such as 'academics understand (environment problems) better'.[2] A variation on this theme is that a rejection of consumerism and the striving for voluntary simplicity is mainly a matter for the well-educated (Durning, 1992, pp. 139–142).

Other hypotheses that have been put forward are that environmental commitment and proenvironmental actions can be linked in different ways to *age*. The young are supposed to be particularly environmentally aware and active in environmental movements; however, a closer examination of what people actually do suggests that proenvironmental activities are more the terrain of older people, especially in terms of preserving natural resources (Finger, 1994; Brand, 1997; Österman, 1998). Some researchers argue that this difference is due to the way that different age groups have their own attitudes to tradition and the establishment; others maintain

[1]By proenvironmental activities, we mean activities which are usually seen in the public debate as environmentally friendly. The concept has nothing to do with actual environmental impact.
[2]Quote from an eco-village resident; cf footnote 11.

that people are moulded by the times in which they grew up, and that a person's generation is therefore an important reason for differences in environmental commitment and proenvironmental activities found between today's different age groups (Kempton, Boster and Hatley, 1995, p. 3).

The conclusions drawn by previous studies into *gender differences in environmental awareness* are diverse. However, according to some studies there is a tendency for women's environmental awareness to be stronger than men's, and that this is directed more at a local level, everyday practice and the conditions under which animals and people live (Buttel, 1987; Lindén, 1994; Merchant, 1992, pp. 183–209; Stern et al., 1993; Brand, 1997). Although women buy more environmentally friendly products, they are no more prepared to make sacrifices and are not noted for their activism (Dietz, Stern and Guagnano, 1998). Much previous research into women's and men's relationships with the environment has been based on the notion that women have an innate affinity to mother earth, as opposed to the exploitative relationship with nature typical of men (Merchant, 1992; Mellor, 1997, Bjerén, 1987; Connell, 1990). Yet we do know that contact with nature, mainly as hunting and fishing, seems to play an important role in men's identity, at least for men living in rural areas (Bjerén, 1981; Dahlgren and Lindgren, 1987; Hägg, 1993; Waara, 1996; Juntti-Henriksson, 1996; Pedersen, 1993; Pedersen, 1995).

Research has shown that the reasons for households acting as they do cannot be reduced to external and internal material conditions (Pahl, 1984; Thompson, Ellis and Wildavsky, 1990). Individuals and households handle similar material conditions in different ways; they have different values, beliefs and attitudes, they define their resources and needs differently and they try to fashion their everyday practice out of different motives. Not everyone has the same ideas about when and where it is important or easy/difficult to do something for the environment (Mårtensson and Pettersson, 2002). We must therefore study a broader everyday life context – all or large parts of everyday practice and the meanings people ascribe to this practice. There is much to suggest that the environment forms part of our everyday life context in different ways, whereby the strength of commitment to the environment varies with area or type of activity, depending on what it means to the individuals and households in question.

The embeddedness of environmental concern in different everyday life contexts, including different meaning contexts, can be a reason for why many traditional explanatory variables have yielded so little. Intra-household interactions also muddy the picture. Values are formed by social interaction and as a result of negotiation and compromise, something which emerges once a holistic perspective on everyday life is taken.

If such broader contexts are of interest, then practice, interaction and meaning in the household's everyday life are important. One such context, which can be linked to environmental commitment and/or proenvironmentalism in everyday practice, is the household's 'life course phase' and composition (partly related to age). An adult's environmental concern can, for example, grow when he or she has a child. However, such relationships do not seem to have been a particularly salient feature of previous environmental research in the social sciences. Our hypothesis is that a household member's interest in and capacity for proenvironmental orientation in everyday practice can vary with transitions and phases in the life course (the phase of life a person is in at any one time), and not least with the household's composition and the presence of children and their ages (Læssøe, Hansen and Søgaard Jørgensen, 1995, p. 71).

Another context is gender order, in the sense of the socio-culturally constructed relations patterns between women and men. Gender order, which covers the sexual division of labour, access to resources and power, and emotional relationships, is a dimension that has a formative impact on a large part of our everyday practice (Thurén, 1996; Thurén, 2000; Connell 1987). The sexual division of labour is particularly important in environmental contexts insofar as solutions to environmental problems are to be found in changed everyday practice.

Household Cultures

To understand why a household acts as it does, we therefore have to acquire an understanding of its everyday life as a whole – including the meaning that the household ascribes to its everyday practice. It is not just environmental attitudes that we need to know, but also a broader structure of values, beliefs and motives that are related to multiple components of everyday practice. One theory that attempts to understand a broader context, while reducing the number of conceivable links between practice and meaning in everyday life, is 'cultural theory' – a theory which tries to interconnect material conditions, everyday practice, social relationships and meaning; it also makes explicit assertions about beliefs about nature (Højrup, 1983a; Højrup, 1983b; Bourdieu, 1979; Bourdieu, 1980b; Douglas, 1982; Thompson, Ellis and Wildavsky, 1990).

According to cultural theory, there are five ways of life, distinguishable by different combinations of *forms of solidarity*, *cultural bias* and *behavioural strategies*. Forms of solidarity are based on a) how strong the bonds are between different individuals and b) on how restricted the individual is by social norms. Cultural bias comprises, amongst other things, ideas of fairness, attitude to authority and myths of nature. Behavioural strategies cover a number of areas, including everyday household practice, or *household culture*. The idea is that different conceptions of resources and needs in general also determine (in different ways) how resources and needs in everyday practice are managed – in 'making ends meet', for example, and in the organization of the domestic daily life. The idea is thus that there are clear links between practice and meaning in everyday life, forms of solidarity and cultural bias.

In the *household culture of the fatalist*, needs and resources are seen as beyond influence. There is little structure and control in the household's everyday practice and finances. Social relations are characterized by the absence of a feeling of group solidarity and society is seen to be largely inaccessible and unequal. People see obstacles to changing or improving their lives, and there is no point in planning or co-operating with others; success, if it happens, is simply a matter of luck. Fatalists do not consider fairness as part of their lives, and nature is seen as capricious and unyielding. We can expect fatalists to think it pointless to do anything for the environment because, in that they lack control over both resources and needs, feel no strong group solidarity, perceive society as being largely closed, see fairness as unrealistic and nature as capricious. As a result they probably consider proenvironmental action as both meaningless and impossible. Any proenvironmental behaviour is either simply a matter of chance or, as tallies with certain previous studies, varies unpredictably. Nevertheless, we expect fatalists to exhibit a low degree of environmental orientation.

The characteristic feature of the *household culture of the hierarchist* is that needs are defined on the basis of collectively determined criteria, such as level on the social ladder. Resources can be collectively influenced – up or down – according to the needs of the group. Everyday practice is governed by structure and tradition. Society is seen as being divided up into groups or strata, the boundaries of which can or should not be crossed, and solidarity exists with one's own group. Fairness means equality before the law. Nature is seen as stable, albeit within certain limits as determined by experts. In the household culture of the hierarchist, we can assume that attitudes towards proenvironmentalism are dependent on the advice and instruction of experts. Since nature is resilient but within limits, there is room for behaviours that are controlled by other considerations than environmental, but when and where a proenvironmental action should be taken is best defined by experts, the behaviour itself being governed preferably by rules and regulations. Since hierarchists believe that resources should be modified to the collectively-defined needs, proenvironmentalism can also mean that different groups should contribute to environmental improvement from different directions. Hierarchists can be expected to adopt an intermediate position on questions of environmental orientation.

In the *household culture of the egalitarian*, everyone is expected to modify their needs to the resources available. Resources tend to be defined as modest and should be shared equally by all. Voluntary simplicity is a common approach to life, as is an association with groups that prioritize equality, including gender equality. Equality and social fairness are the guiding ideals of everyday practice. Rejecting the hierarchies of society, egalitarians consider themselves as belonging to a collective in which no one person is of more value than the next. Fairness is defined as equality of outcome. Since nature is seen as fragile, the demands of the egalitarian household involve concerted proenvironmental action. Everything that reduces the consumption of the earth's limited resources and protects our fragile environment has to be done. Everyone is expected to contribute equally to caring for the environment. Since needs can be modified downward, it should be easy to behave proenvironmentally. Such behavioural adaptation is not so much a sacrifice on behalf of the environment than a way of expressing an allegiance to a lifestyle of equality and the rejection of individualism and competitiveness. The egalitarian can be expected to have the highest degree of environmental orientation.

The *household culture of the individualist* is flexible in its attitude towards resources and needs, although efforts are made to increase both. Cleverness and a high propensity to take risks are to be rewarded. There is a relatively high level of adaptability when it comes to making material ends meet. The organizing of everyday practice is characterized by flexibility; group affiliation and solidarity are inconstant and society is seen as open to mobility. Fairness is a matter of the equality of basic opportunities. Nature, perceived as benign and manipulable, is also seen as flexible. People with an individualistic way of life can be assumed to act proenvironmentally as long as it does not intrude on their individual freedom. The solution to environmental problems lies in establishing a system of private ownership, since this would involve everyone bearing the cost of their own actions. The art of making people act proenvironmentally entails giving them incentives to do so and if this is done, there will be environmental improvement. Unlike hierarchists, individualists do not think that experts know best. Individualists also claim that there is a vast creative potential in people, and that this should not be confined by rules and prohibitions. Since nature is robust, it can also withstand unsuccessful experimentation. The level of environmental orientation can be expected to be low amongst individualists.

In the *household culture of the hermit* too, resources and needs are seen as open to influence, although in this case the strategy is to find a point of equilibrium between relatively minor needs and sufficient resources, and to achieve a state of independence.[3] In many respects, this household culture is a cross between the other four. This also applies to their own breed of social solidarity: 'hermits' do not affiliate themselves with groups in any fully committed way, but spend time with the people they enjoy being with. Man is seen as part of nature. However, it should be pointed out here that the ideas put forward by cultural theory about the hermit's views of nature and fairness are undeveloped. The household culture of the hermit should involve practices and beliefs concerning a proenvironmental orientation that are independent of how others think and act. According to the theory, there is a sense of affinity with nature, which can mean that a proenvironmental orientation

---

[3]The hermit category is a later addition to this theory.

is considered as natural and effortless. Since the hermit's way of life is a blend of, and a cross between, the other four, we can assume that the proenvironmental orientation of the hermit lies at an average position.

In summary then, we can expect the lowest level of environmental orientation to be found amongst fatalists and individualists, although for different reasons; a mid-way position to be adopted by hierarchists and hermits, again for different reasons; and finally, the highest level to be the preserve of the egalitarian.

A Survey on Everyday Life and the Environment in Stockholm

In a recently completed survey of 634 middle-aged households in the Greater Stockholm area, attempts were made to quantitavely test these ideas and examine whether household culture is indeed linked to environmental orientation.[4]

According to cultural theory, household culture is not determined by socio-economic conditions and other material factors; in other words that an individual's view of resources and needs is not governed by objectively quantifiable resources. An important question is whether this is the case in our study as well, whether household cultures are independant of material conditions. We will therefore begin by describing the household cultures as we have assessed them and how they relate to material conditions.

Household Cultures

By household culture is meant here primarily the interviewee's own opinions about whether or not the following are true for their own household: We/I ...

- ... find it difficult to save money; ... have few long term plans (fatalist household culture);
- ... have fixed meal times; think it is important to carry on family traditions (hierarchist);
- ... like to plan carefully so that financial risks are not taken; hate the use of credit and try to never use it (hermit);
- ... like to look successful; are very fashion conscious (individualist);
- ... avoid products made by oppressive institutions; never celebrate holidays that are created simply out of commercial interest (egalitarian).
  (Dake and Thompson, 1993).

[4]The Stockholm survey involved interviews with a probability sample of 634 households (ages 35–54) in the Greater Stockholm region (except peripheries). Paul Fuehrer has also participated in this work.

Two additional statements are not correlated to household cultures in the predicted way in our study: 'often buy second-hand furniture' (theoretically fatalist and/or egalitarian) is not clearly related to any household culture, and 'think it is important to be on time' (theoretically hierarchist) is mainly part of the household culture of the hermit.

Table 1   Factor analysis: patterns of household cultures (varimax rotation)*
634 households

In our household we ...	Household culture of the ...					*Commu-nality*
	hermit	hierar-chist	indi-vidualist	fatalist	egal-itarian	
... hate the use of credit and try to never use it	75		−11			58
... like to plan carefully so that financial risks are not taken	70			−33		61
... think it is important to be on time	63	27				48
... think it is important to carry on family traditions		69	12			49
... have fixed meal times	17	61				40
... like to look successful			82			69
... are very fashion conscious		36	67			58
... often purchase second-hand furniture		−37	37	28	34	47
... have few long term plans			−16	79		65
... find it difficult to save money	−18		23	68		56
... avoid products made by oppressive institutions		36			77	73
... never celebrate holidays that are created simply out of commercial interest	11	−30			66	54
Variance	*16*	*13*	*10*	*9*	*8*	*56*

*The interviewee's own opinions about whether or not the statements in the table are true for their own household. Responses were indicated on a 5-point scale, from 1. 'disagree strongly' to 5. 'agree strongly'.
Factor loadings have been multiplied by 100. Only factor loadings >10 are printed. Eigenvalue=1.

Despite the limited nature of our quantitative assessment, the five different household cultures were discernable in a factor analysis and can therefore be used in our analysis of proenvironmental actions.

Before we look into the relationship between the different household cultures and 'doing something for the environment' however, we will discuss their relationships with a number of important material variables. The first variable is 'life course phase': 'younger' and 'older' single households and childless couples (the adult in a single household/the woman in a couple are aged 35–44 and 45–54 respectively), households consisting of two adults and children of different ages, and single parent households. The second variable is the sex of the interviewee. The third set of variables is 'socio-economic status', assessed primarily on the basis of the highest level of education in the household and the average adult income.

An examination of the mutually independent bearing that life course phase, gender, educational background and income each have on variations in household culture, shows that as explanatory factors, they are of negligible importance. However, these background factors have a varying degree of importance for the different household cultures: fatalistic and hierarchical household cultures are most readily linked to background factors, while the weakest links are exhibited by the individualist and the egalitarian.

Fatalistic household cultures were mainly found amongst single parent households (child plus single adult – usually the mother) and households on low income. Fatalist young couples were comparatively rare.

The hierarchical household culture was relatively common amongst our households with children and childless couples over 45, and it was more characteristic of women than men. The over-representation of women and families with children is probably related to our indicators of the hierarchical household culture which are largely about family tradition and everyday routine, things which are likely to be important to parents – and then particularly for women, who more often than not have the practical responsibility for the children's day-to-day lives. Socio-economic conditions also have a certain bearing: a low level of education but a relatively high income are also associated with the hierarchical household culture.

The hermit household culture was largely the territory of low income earners and women. It was also linked to life course phase, in that it was also found amongst childless couples over 45.

Finally, individualistic and egalitarian household cultures had less of a correlation with the household's external and internal material conditions. However, one noticeable phenomenon was that while an individualistic household culture was rare amongst older couples, it was noted, to a certain extent, amongst the well-educated.

Egalitarianism was a feature of the well-educated, low income households. Even though our method of assessment may be inexact, this suggests that it is

Table 2   Regression of household cultures on background variables. Standardized coefficients. 634 households

	Household culture of the ...				
	fatalist	hierarchist	hermit	individualist	egalitarian
Younger couple. no children[a]	−.112*	−.003	.055	−.093	.005
Couple + youngest child under 7[a]	.066	.266**	.020	−.071	−.015
Couple + youngest child aged 7–14[a]	.006	.296**	.052	−.155*	.048
Single parent households[a]	.192**	.126*	.045	.060	.042
Couple + youngest child over 14[a]	−.083	.232**	.057	−.063	.011
Older couple. no children[a]	−.053	.149**	.120*	−.200**	.111
Older single households[a]	.050	−.063	.092	−.080	.057
Woman (0) / man (1)	−.055	−.108**	−.136**	.018	−.078
Highest educational level	−.021	−.142**	−.063	.096*	.116*
Household income	−.179**	.092*	−.155**	.036	−.113*
R^2	*12.3*	*12.6*	*7.8*	*5.6*	*3.4*

'Younger'=age 35–44; 'older'=age 45–54
[a]Reference category: younger single households
*5% confidence level; **=1% confidence level

mainly households with high cultural capital but low economic capital that belong to the egalitarian culture. According to Pierre Bourdieu, people with this capital combination exhibit a 'distinctive taste' based on a rejection of established tastes and conspicuous consumption but preferences for voluntary simplicity and the agenda of alternative movements (Bourdieu, 1979; Bourdieu, 1980b).

An additional consideration is that most households are unequally distributed between four residential categories reflecting different degrees of population density: detached housing, terraced housing, suburban apartment blocks and inner city.[5] A comparatively large number of households with a fatalistic culture are found in suburban apartment blocks, hierarchical and hermit household cultures mainly in areas of detached housing, with the egalitarian household culture being more representative of inner city than anywhere else. The individualistic household culture cannot be associated with any particular residential category.

There is, therefore, a certain correlation between household culture and other circumstances and conditions in the life of the household. We will therefore examine what bearing the household cultures have on environmental orientation

[5]Results based on variance analysis (Tukey's Studentised Range).

compared to these material circumstances and conditions. We will then discuss two of these factors in more detail – life course phase and gender order, two factors that have been identified as components or dimensions of the everyday life context. Henceforth, the main issue is how different kinds of environmental orientation are embedded in the everyday life context, and, more specifically, how it relates to household culture.

Household Culture as Everyday Life Context

In previous studies, interest in cultural theory has mainly focused on the connection between 'cultural bias' (such as conceptions of fairness and nature myths) on the one hand, and the perception of environmental risks and environmental attitudes on the other (Grendstad, 2001; Ellis and Thompson, 1997; Steg and Sievers, 2000). As for how the environment enters the everyday life context which the different household cultures encapsulate, fewer studies are available. Ideas of household culture are a late addition to the theory, and have not yet prompted that many empirical studies. However, two qualitative and one quantitative study of household cultures have been carried out in England (Mars and Mars, no date; Dake, Thompson and Neff, 1994; Dake and Thompson, 1993).

We will now discuss three proenvironmental activities: recycling, the purchase of organic food, and car travel. All of these are considered in the public debate to be important in an environmental context and can be seen as expressions of a willingness to do something for the environment. Unless otherwise stated, the results below are based on an analysis of the separate impact that the different explanatory factors have on the variables under investigation.

Recycling

Recycling is an activity that is neither expensive nor especially time consuming – in fact, in certain parts of town it is even financially beneficial and practical. It is also widely regarded as environmentally friendly.

According to earlier studies on recycling activities, the different properties of the recycling system's design and organization are just as important as environmental attitude (Derksen and Gartrell, 1993; Berger, 1997; Mårtensson and Pettersson, 2002). A Swedish survey reveals that there are also differences between groups in recycling. Women demonstrate a higher degree of recycling than men, middle aged people a higher degree than the young. Combining age and gender shows that men under 30 are a surprisingly low-performance group, while women between 50 and 64 are particularly active (Förpackningsinsamlingen, SIFO, 1999). A certain correlation with socio-economic status can also be observed: high-level white-collar workers recycle most, blue-collar workers least.

In our study of middle-aged households in Stockholm, it is common to recycle, and most people say that they recycle at least some of the seven most commonly studied materials.

Assessing the importance of household culture on recycling – compared with the properties of both residence and residential area, the household's socio-

Table 3 **Regression of recycling (7 most common items x frequency, 0–28) on background variables, household cultures, environmental attitude index and attitudes related to recycling. Standardized coefficients. 634 households**

| | Model | | | | | |
	A	B	C	D	E	F
Younger couple, no children[a]	.020	.004	−.014	−.025	−.038	−.045
Couple + youngest child under 7[a]	.161*	.099	.103	.081	.061	.050
Couple + youngest child aged 7–14[a]	.204**	.117	.103	.077	.063	.047
Single parent households[a]	.072	.057	.062	.052	.044	.038
Couple + youngest child over 14[a]	.198**	.129*	.116*	.101	.097	.088
Older couple, no children[a]	.159**	.114*	.074	.072	.054	.054
Older single households[a]	−.011	−.021	−.039	−.039	−.037	−.039
Woman (0) / man (1)	−.143**	−.151**	−.123**	−.109**	−.071	−.065
Highest educational level	.045	.063	.059	.059	.033	.033
Household income	.045	.029	.054	.033	.059	.043
Population density of residential area (4 categories: low–high)		−.221**	−.195**	−.203**	−.196**	−.201**
Fatalist (factor score)			−.072	−.064	−.066	−.061
Hierarchist (factor score)			.016	.024	.013	.019
Hermit (factor score)			.138**	.136**	.124**	.125**
Individualist (factor score)			−.043	−.027	−.048	−.036
Egalitarian (factor score)			.127**	.092*	.104**	.080
Environmental attitude index (1–18)				.152**———		.117**
Recycling is important (1–4)					.184**	.159**
Confidence in recycling system (1–4)					.116**	.113**
R^2	7.7	12.0	15.8	17.9	20.4	21.5

'Younger'=age 35–44; 'older'=age 45–54
[a]Reference category: younger single households
*5% confidence level; **=1% confidence level

economic circumstances, life course phase and gender – it appears that differences in recycling practice are mostly attributable to residential category. Recycling tended to be most frequent in areas of detached housing, lower in areas of suburban apartment blocks, and lowest of all in the inner city. Besides this, there were two household cultures that were intimately linked to recycling behaviour – the hermit and egalitarian. Another important factor was gender, with women professing to recycle more than men. Life course phase had little significance in this model because nuclear families dominate the detached housing areas; when residential category is eliminated, the life course phase factor increases in importance. We can also add that the household's socio-economic circumstances (level of education and income) had no significance in this context.

When the analysis also includes the responses to two attitude questions of direct relevance to recycling (to what extent does the interviewee think that it is important to recycle, and to what extent is he/she confident that the material is handled as it is meant to be), the hermit and egalitarian household cultures retain their level of significance. It also appears that these two attitude factors are important, the former slightly more so than the latter. It should, however, be mentioned that it was mainly people from an egalitarian household culture who thought that recycling was important and who expressed confidence in the system. When these attitudes are included in the analysis, residential category is still, however, the factor with the strongest impact.

An important conclusion that can be drawn from all this is that both household culture and attitude to recycling are important, but equally so is the presence of an efficient recycling infrastructure. Our results confirm previous studies in indicating the significance of this. To understand differences in recycling frequency, we should extend the everyday life context to include certain properties of the residence/ residential area and infrastructure: the household's place in the urban structure and the differences in infrastructure that exist between different types of residence and residential area. It seems to be largely a matter of how the recycling infrastructure is organized – how close, practical, aesthetically pleasing it is and the rules that apply in different parts of the city and its environs.

The Purchase of Organic (KRAV) Food[6]

Another proenvironmental activity (i.e. one that is often associated with environmental friendliness), is the buying of organic food, although organic or KRAV labelled foodstuffs can be bought for other reasons too: health, for instance, which is seen as having a particularly intimate relationship to the environment (Kempton, Boster

---

[6]KRAV is a certification organization for the organic production of foodstuffs and consumer goods.

Table 4 **Regression of purchase of organic (KRAV) food (number of food categories x freqency, 0–52) on background variables, household cultures and environmental attitude index. Standardized coefficients. 634 households**

	Model		
	A	B	C
Younger couple. no children	−.014	−.021	−.039
Couple + youngest child under 7	.094	.089	.055
Couple + youngest child aged 7–14	−.043	−.065	−.106
Single parent households	−.075	−.074	−.089
Couple + youngest child over 14	−.045	−.060	−.081
Older couple. no children	−.029	−.058	−.059
Older single households	−.076	−.079	−.078
Woman (0) / man (1)	−.158**	−.147**	−.123**
Highest educational level	.056	.041	.040
Household income	.036	.038	.003
Fatalist (factor score)		−.050	−.038
Hierarchist (factor score)		.038	.054
Hermit (factor score)		−.039	−.039
Individualist (factor score)		−.028	−.003
Egalitarian (factor score)		.173**	.112**
Environmental attitude index (1–18)			.260**
R^2	5.2	8.7	14.7

'Younger'=age 35–44; 'older'=age 45–54
[a]Reference category: younger single households
*5% confidence level; **=1% confidence level

and Hatley, 1995; Douglas, Gasper, Ney and Thompson, 1998). KRAV food often costs more, but can be bought in most parts of the city.

Earlier studies have demonstrated different correlation patterns regarding the groups that buy organic food. According to one national study in Sweden, women between the ages of 50–70 were a high-consumption group (Holmberg, 1999). Households with childless couples also ranked high. Differences in educational background between frequent and less-frequent buyers of organic food were not apparent, although people with an educational background up to lower secondary level were highly represented amongst the frequent buyers – a phenomenon which is probably related to age.

In the Stockholm survey, the interviewees were asked how often the household bought products from the KRAV-labelled range. More than half answered that they did so sometimes.

When we examine how the amount of organic food purchased varies with household culture compared to the household's level of educational background and income, life course phase and gender, it transpires that egalitarian household cultures account for most of the variation in KRAV purchasing behaviour. Gender was also important: women claimed to buy more than men. One interesting finding is that income did not lead to any differences in purchasing behaviour despite the fact that KRAV food is more expensive than conventionally produced food.

When an environmental attitude index is brought into the analysis, the explanatory value of the egalitarian household drops somewhat, which is consistent with a close positive correlation between egalitarianism and environmental commitment.[7] Nevertheless, the egalitarian household culture remains a significant explanatory factor. In a similar way, KRAV purchases continue to be strongly linked to women. When it comes to the purchase of organic food, therefore, it is both the broadest appraisal of everyday life context (household culture), gender and attitude which have bearing on KRAV purchases.

Car Travel

It is widely accepted that the car is not environmentally friendly. We can assume that car travel is largely dependent on exterior material circumstances, mainly category of residential area and transport structure.

Previous studies have shown that car ownership and car travel is most widespread amongst men, and that it increases with income, both of the individual and the household. Category of residential area and access to public transport are also key factors – car travel per person is significantly lower in inner city areas than in more sparsely populated areas (Svesnkarnas resor, 2000).

In the Stockholm study car travel was measured as mean number of kilometres traveled by the adult(s) of the household.

The extent of car travel for adults in middle-aged household in the Greater Stockholm area has no relationship with household culture.[8] Not even in egalitarian households is there any less car travel. It appears that the amount of car travel is

---

[7] Our environmental attitude index is composed of answers to several questions concerning among other things preparedness to dedicate more time and/or money to changing everyday life towards a more proenvironmental orientation, proenvironmentalism as a motive for different household activities.

[8] Includes business travel.

Table 5 **Regression of car travel (mean number of kilometres per adult/ year) on background variables, household cultures and environmental attitude index. Standardized coefficients. 634 households**

	Model			
	A	B	C	D
Younger couple. no children[a]	.020	.010	−.008	−.009
Couple + youngest child under 7[a]	.042	.001	−.034	−.035
Couple + youngest child aged 7–14[a]	.091	.034	−.023	−.028
Single parent households[a]	.095	.085	.056	.054
Couple + youngest child over 14[a]	.091	.046	.015	.014
Older couple. no children[a]	.087	.057	.033	.028
Older single households[a]	−.003	.046	−.028	−.033
Woman (0) / man (1)	.152**	.147**	.148**	.144**
Highest educational level	−.116**	−.104*	−.105*	−.105*
Household income	.219**	.208**	.202**	.207**
Population density of residential area (4 categories: low–high)		−.147**	−.146**	−.144**
Fatalist (factor score)			−.006	−.008
Hierarchist (factor score)			.026	.024
Hermit (factor score)			−.006	−.005
Individualist (factor score)			.038	.034
Egalitarian (factor score)			−.063	.006
Environmental attitude index (1–18)				−.038
R^2	7.7	9.6	9.8	9.9

'Younger'=age 35–44; 'older'=age 45–54
[a]Reference category: younger single households
*5% confidence level; **=1% confidence level

primarily linked to income, gender and residential category. The more well-off a person, the more car travel he or she is engaged in. This can be compared with the finding that car travel had a negative correlation with educational background.[9] Not surprisingly, women use cars, either as drivers or passengers, to a much lesser extent than men did. Furthermore, most car travel is observed within detached housing areas and least within the inner city: another well-known fact.

[9]i.e. capital combination.

Another finding is that people from environmentally committed households do not travel by car any less than those from environmentally indifferent households, judging by the explanatory values that our environmental attitude index applies to car travel compared with the explanatory factors above. This can be compared with the finding that two-thirds of those who used their own car consider cars to be damaging to the environment. The reasons for using a car obviously outweigh any reasons there might be for not using it. A wide range of different studies shows that the car makes everyday life easier and suits the drivers' many and varied interests and motives. Bearing this in mind, it is hardly surprising that neither environmental commitment nor household culture has any relevance to the extent of car travel (Hagman, 2000).

In terms of car travel, the environment is of little relative importance compared to everyday practice and other meanings and motives. The context here is a combination of the household's financial status, gender order, the properties of the residential area and urban structure.

Limited Everyday Life Contexts

Besides household culture, there are two more limited contexts, or components of everyday life, which are of particular interest when trying to understand how a household's proenvironmental orientation is embedded in everyday life contexts: the household's life course phase and gender order.

Life Course Phase

During a person's or household's life time, changes take place between life course phases, changes that are linked to turning points or more gradual transitions within certain spheres such as education, work and the family.[10] As suggested above, certain differences of life course phases in everyday practice and meaning are particularly interesting in terms of environmental orientation.

Previous studies suggest that young people protest against 'the rat race' and often become involved in different brands of activism. At the same time they are usually considered as having an experimental lifestyle and to be in search of an identity. They are also often regarded as being particularly consumption orientated. Because of their egocentrism and consumption requirements, young people are not expected to have any great interest in any applied environmental orientation in their everyday practice, no matter how strong their environmental

---

[10]Changes from one life course phase to another are not necessarily linked to age-related changes. A person can, for instance, build a family more than once, and starting a family for the second time can mean something completely different than the first.

commitment (Finger, 1994; Brand, 1997; Österman, 1998; Kempton, Boster and Hatley, 1995).

When people start a family, there is a tendency for their everyday practice to become more routine. Parenthood seems to generate a particular interest in the health consequences of their choice of housing, food and other products, and this is often associated with environmental friendliness. It can also mean that parents become involved in local political issues in order to improve their children's situation (Sorbom, 1999). Then, once the children reach school age, other tendencies emerge. Their children now come home from school, admonishing their parents and pushing them towards a more proenvironmental orientation (Martensson and Pettersson, 2002). On the other hand, it appears that the propensity for choosing environmentally friendly alternatives in households containing older children and teenagers is inversely proportional to the child's consumption demands. When there are no longer any children remaining in the household, the priorities may change. An improved financial situation for the household and more free time from work can make room for other activities and other consumption activities. One noticeable tendency is that old people, like the youngest, display signs of protest against a single-minded orientation towards material things as well as an increased interest in 'soft' values (Finger, 1994; Brand, 1997; Osterman, 1998; Kempton, Boster and Hatley, 1995). Care for their grandchildren and worry about their future are other concerns expressed by older people (Mårtensson and Pettersson, 2002).

Gender Order

Gender order, meaning the historically constructed patterns of relations between women and men and the cultural beliefs about these patterns, is a fundamental element of everyday life context. Three structures can be distinguished within a society's gender order: firstly, there is the gender-based division of labour; secondly, each gender's own access to resources; and finally, the emotional relationships between the sexes (Thurén, 2000; Connell, 1987). The three structures can be studied in terms of their range (how many areas are gender-determined – 'gendered'), force (how strongly gender is symbolized in actions, body language etc., and the extent to which deviations from the cultural norms are sanctioned) and hierarchy (superiority/inferiority). Particularly interesting is the gender-division of domestic chores, since the environmental orientation being studied is related to activities in the home.

Gender specialization in the home, with the woman responsible for the caring and the man both responsible for the providing and in need of being cared for, has declined in Western countries. The equality ideal can be said to have acquired an ideological hegemony. But despite this, women's and men's areas of responsibility when it comes to domestic chores display a striking resilience. The woman still has the main responsibility for all work related to food, cleaning, textiles and care, and she performs more house-work tasks than the man. The man is responsible for building, technical maintenance and transport and is generally the one responsible for provision, while still maintaining a position of nurture-dependency vis-à-vis the woman (Bjerén, 1991; Hagg, 1993; Back-Wiklund, no date; Rydenstam, 1992). One general tendency highlighted by gender research is that in the interaction between spouses/partners, it is the woman who tends to take responsibility for the well-being of the entire family, which means, in effect, taking on an extra job (Haavind, 1984; Haavind, 1982; Haavind, 1992). One theory is that women reason more often than men on the basis of a 'rationality of caring', meaning that they think in terms of broad social relations, and in concrete situations consider the consequences for the many; men, on the other hand, reason from a value hierarchy and of the people involved, take into consideration only those to whom they are closest (Gilligan, 1982; Sandberg, 1998). Studies also show that a caring responsibility can lead to feelings of guilt in women (Elvin-Nowak, 1999). There is much to suggest that women also take responsibility for the household's environmental orientation (Plumwood, 1993; Buckingham-Hatfield, 2000).

Let us now discuss the impact that life course phase and gender order have on our examples of different environmental activities.

The Recycling Example

Household culture has an important bearing on variations in the degree of recycling activity. However, our results concerning the importance of explanatory factors for these variations also clearly illustrate the role of gender order. Recycling is an activity which is intimately linked to the domestic kitchen surroundings, typically the woman's territory, and women appear to take on much of the extra work that recycling involves. An interesting hypothesis in relation to this is that greater equality in practice – i.e. men taking more of an active role in the household work – leads to the household dedicating less total time to household production, and thereby generating more waste (Godbey, Lifset and Robinson, 1998). It cannot be taken for granted, however, that men would also take on more of the domestic recycling workload.

Life course phase is not a decisive recycling factor in our study. If we eliminate category of residential area from the analysis, nuclear family households emerge as active recyclers. Most nuclear family households live, however, in detached housing areas, and the properties of the residential area supersede life course phase/type of household. One aspect of the way of life in the detached residential area seems to be the possession of a wide range of different material goods (large volumes of household capital goods, leisure equipment and vehicles). Large households and abundant possessions create more waste. This can be compared with the fact that nuclear family households in the inner city have a lower material usage (measured by weight). The difference between the nuclear family households of these two residential categories suggests that those who can, choose to settle in the type of area that for one reason or another best suits their needs. One such reason might be what is embodied in household cultures – we have seen that the household cultures are unevenly distributed over the different residential areas.

The Purchase of Organic Foods Example

Food is an essential aspect of people's everyday lives. Many beliefs and values are linked to food, and it enters our social relationships in different ways. Constant negotiations take place in families on food and meal times, and choosing food to suit all household members is a task for which many different considerations have to be adapted to each other. In households with children, the adults' choice of food is an expression of concern for their children's health, although this can clash with another concern, that of satisfying their children's taste (Warde, 1997). In our Stockholm survey, households with children purchased no more KRAV foods than others, although a difference was noted in the reasons stated for doing so. There was a marked tendency towards 'appearance' or 'taste' being important motives for choice in households with children over the age of 7; this can be compared to households with children under 7, where important motives were connected to health and the environment. Life course phase is thus an important component in the everyday life context which involves food. When people have children, a drastic transformation usually takes place which, for the parents, has both a short and long-term impact on their everyday practice and priorities. Concern for the child's health and well-being is a particularly important factor (Mårtensson and Pettersson, 2002).

Women have a key part to play in the food area; the choice and preparation of food and the social relations associated, for example, with what appears on the table and the actual eating of the food at home (De Vault, 1991; Beardsworth and Keal, 1992). The actual gender division of labour (including the woman's traditional role as carer) as well as the values associated with gender order impact on the food area, and thus the choice of KRAV food. Gender order is an important factor in our study, although it is clear that the volume of organic foodstuffs varies within this context (Kalof, Dietz, Stern and Guagnano, 1999).

The Car Travel Example

As is the case with recycling and the purchase of organic food, car travel is gender related, and, again like recycling, it is linked to category of residential area. However, unlike the other two forms of proenvironmental action, car travel lacks any relationship with household culture and environmental attitude.

We know from previous studies that practical considerations such as time and convenience play a major role in the choice of transport. Life course phase is of no importance here; true, nuclear family households may be over-represented in areas where the car can be thought to fulfil a vital practical function, but other factors are more important. Earlier research demonstrates that gender order is also a central factor of daily travel (Hjorthol, 1998; Hjorthol, 2001; Berge, 1998; Hagman, 2000). Many of these studies provide examples of the car's inherent identity-building potential, something which tends to be more important for men.

Car travel is also special in that while it has a positive correlation with income, it has a certain negative correlation with educational background. Those who exhibited low car-usage tended to be those with low financial capital and high cultural capital (education capital) (Bourdieu, 1979; Bourdieu, 1980b).

Narratives about Life Course Experiences in Swedish Households

So far, we have been discussing contexts in which individuals are captured in a certain phase of a their adult life. However, there is reason for broadening the life course perspective to examine the history of experiences that the individual carries with her/him. This is an everyday life context that has a potential bearing on a person's attitudes to the environment, although this is an idea that may very well contradict theories of the post-modern society, which assert that an individual's life, now more than ever, is influenced by circumstances beyond his or her immediate surroundings. It is claimed that a person's way of life is shaped less by immediate and direct experiences and more by mediated impressions. This seems to be confirmed in the environmental field, where access to information via the media is sometimes given a decisive role in the awareness of and concern for both local and global environmental problems. However, it is easy to overestimate the impact of public opinion and the media, bearing in mind that access to information about environmental problems does not satisfactorily explain variations in the environmental orientation of everyday practice. Ideas and beliefs, formed by the public debate, seem to be of rather limited importance when it comes to changing everyday practice (Brand, 1997).

It seems, therefore, that there is still room for influencing factors which are seen to belong to a former society. This applies particularly to the role that childhood experiences play in shaping a person's life. An individual's personal history and life course, and the way she is marked by, and has reacted to, the circumstances of her upbringing can be important in understanding environmental orientation in everyday life (Mannheim, 1972 [1952]; Lunt and Livingstone, 1992; Frønes, 1997). This does not mean that we can expect the reproduction of the behavioural patterns of childhood to have a given relationship with environmental orientation. Experiences can conceivably influence environmental orientation in everyday practice in a number of ways. What is passed on, for example, could equally be childhood experiences of environmental preservation and composting as much as consumerism and environmental indifference.

We will now look at what importance people ascribe to previous life events and experiences for certain aspects of their current everyday practice and relationship to the environment. The emphasis in this study is on how people view the importance of their childhood and parental influences.

There are few previous studies in this field. One such was by Matthias Finger, who applied a biographical life-world approach to a series of interviews and surveys he conducted in Switzerland on how life experiences influence environmental behaviour. His hypothesis was that individuals are embedded in a socio-cultural life-world, which includes important life experiences of the environment (Finger, 1994).

We interviewed 59 people in 34 households in three eco-village groups, collective housing households and relevant control groups.[11] Each interviewee was asked to describe the evolution of their current way of life, and the circumstances that could have influenced their life course. Questions were also asked about their parents' upbringing and way of life during their (the interviewees) childhood. In doing so we sought to obtain a person's thoughts about how important their parents and other key persons were as either positive or negative role models.

Life histories and similar material can be seen as stories, variously modified by the teller according to what is thought might be socially desirable, how the interviewer is perceived and so on (Ehn, 1992; Roos, 1992; Åström, 1986; O'Rand and Krecker, 1990; Giele and Elder, 1998). The extent of this problem cannot be determined, but

[11]The aim of the study as a whole was to investigate livelihood, organization of everyday life and degree of ecological orientation in everyday practices and environmental concern in 129 households chosen from nine groups representing different degrees of environmental friendliness in the individual dwelling and the neighbourhood infrastructure. The sample contains households in two 'eco-villages', 1) one in a locality in a rural area in Northern Sweden, 2) the other in a suburb of a medium-sized town in Middle Sweden, furthermore 3) households planning – and planning to move to – an ecovillage in a Stockholm suburb, 4) households in a block provided with some ecological infrastructure in central Stockholm and 5) households living in a collective housing unit in a Stockholm suburb. All groups have their control groups in conventional housing in the vicinity except the households planning an ecovillage. The study comprised three steps: one basic interview concerning mainly livelihood and organisation of everyday life in 129 households, a mail questionnaire mapping the informal help network of a subsample of 34 households, a semi-structured in-depth interview in the same 34 households concerning the relationship between practice and values and also a section on life course experiences.

it must be assumed that the interviewees, to differing degrees and more or less consciously, tend to reinterpret their lives. It might be expected, for instance, that environmental commitment and other socially desirable motives are given special emphasis at the expense of other motives. However, it is interesting to note that a number of interviewees made no attempt to put themselves in a more favourable light and stressed the fact that aspects of their way of life which could be seen as environmentally friendly were based on quite other reasons.

Looking at the main tendencies, it is possible to detect different kinds of influence on environmental orientation. The *first* is that of previous experiences or role models. Such influences are found in five areas:

- the keeping and making use of natural resources and items;
- thrift in general (as opposed to consumerism and a use-and-dispose mentality);
- food cultivation;
- an interest in the outdoor life in general;
- the influence of those a person considers pioneers and 'good examples' in environmental matters.

The *second* type of influence is rooted in changes in a person's life course. Important changes here include starting work, marriage and above all, parenthood. The arrival of grandchildren is also important. Of interest here are experiences linked to transitions and turning points in a person's life course during adulthood.

The *third* kind of influence has a more indirect impact on a person's life – namely public events and the socio-cultural circumstances of society in general. These include times of hardship and war, economic growth, the emergence of the consumer society, the transition from the work-ethic to the leisure-ethic, the general 'zeitgeist' and the public debate on environmental problems.

Nearly all interviewees saw some kind of pro-environmental influence from *previous experiences or role models*. Two-thirds of this group mentioned such an influence as having existed during their childhood. An important area for such an influence is keeping and making use of natural resources. Nearly 75 per cent of the interviewees had, during childhood, at least one parent who made efforts to keep and make use of available natural resources, to combat waste and to maintain things so that they lasted longer. About 75 per cent of the people who had some such experience from their childhood claim to have taken after their parents in this way. One example of how the keeping and making use of natural resources can be passed down from one generation to the next is given by this woman from an eco-village in northern Sweden:

> – If you're taking a walk in the forest, you're supposed to pick up every dry stick you see lying around, father will put it on the fire – nothing's to be done without a purpose. And then you have to pick berries … but you don't have the time.

Interviewer: – But have you inherited any of this?
– Oh yes, a lot.
Husband: – You've got to take care of everything. It's there in the back of your mind the whole time, and when Lena gets old, she'll get just like her mother. It's usually like that. When we then go to the cottage and Lena's parents aren't alive any more and if you go past a stick you pick it up, because that's what they did, and that idea is to be kept alive. And then everyone at home on their sofas will say you're crazy, you've got enough wood to last 20 years.

The attitude to keeping and making use of natural resources does not, however, remain constant over a person's life course. It is not unusual for teenagers to renounce their parent's patterns of behaviour, only to adopt certain aspects of them at a later date. In addition to parental influence, there is also that of elder relatives, which can also help to form a young person's adult behaviour patterns in this respect. Close relatives in their grandparents' generation in particular can be both positive and negative role models when it comes to a person's attitude to money and resources. The antithesis of keeping and making use of natural resources – waste and consumerism – was a less common experience in our households. Of the eight people who grew up in homes which they described as decidedly consumer-oriented, half reacted against this at an early age. Two of these people partly continue on the same lines, while the others did so in their early adulthoods only to renounce what they saw as over-consumerism later on. Those who turned against their parents' consumer-orientation had developed an environmental commitment, expressly associating consumerism with environmental 'unfriendliness'. One form of keeping and making use of natural resources that was considered a formative experience was composting. Nearly all the older interviewees claim to have always composted, despite changing trends. Many younger people, often from rural backgrounds, had seen their parents and others compost during their childhoods and had themselves developed the technique based on this experience.

Food cultivation appears to have been a very widespread activity in the immediate childhood surroundings of the interviewees. Almost all of them had had direct contact with food cultivation at some time in their lives. For most people, this involved garden cultivation for production and pleasure, sometimes with a potato bed. Some drew their experiences of cultivation from agriculture, having either grown up on a farm or having spent summers with farming relatives. Nearly all had happy memories of food cultivation, as often from their childhood homes as from those of their grandparents. Many had been inspired to grow their own food as adults, chiefly those, perhaps, who had been allowed to help or grown things on their own. It appears to have been a significant factor to have 'got their hands dirty'. For some interviewees, these experiences were associated with certain people. Some label one of their parents as having a 'gardening lifestyle'. Of those who grew up with food cultivation and contacts with nature, half say that this made a lasting impression on their own ways of life, and that much of their adult behaviour still remained close to nature. These experiences seem for the most part to have also made the individuals more environmentally active. Some people, of different ages, learnt as children not to litter the countryside, something they point out as having since proved environmentally friendly. Some gave particular emphasis to their parents' interest in the countryside and the outdoor life, which they believed led to their own interest in nature. In cases where food cultivation and other such 'nature' interests experienced during childhood had influenced lifestyle in a proenvironmental way, it was said to have done so gradually.

It was not only the 'nature' experiences of childhood that influenced attitudes to the environment. In some cases, things experienced in adult life also contributed to change – by either reinforcing or substituting childhood experiences. Immediate and concrete personal experiences of environmental destruction could reinforce an old environmental interest, like in this man from an eco-village in the Swedish midlands:

> We were out fishing in a lake in the middle of the forest, in the winter ... All snow on the lake had been rained away, and it was just pure ice and we started fishing. And we said to each other, God the hole's dirty, and there was a film on the water, it's oil ... On the way home we said, imagine it being so dirty, where'd it come from, it's got to come down with the rain and snow. And so far, it was many, 50–60 km from some factories. We saw that there must have been a lot of pollution and everything .... That made us think that there are so many things that aren't good. It affects you.

Ten or so interviewees said that environmentally committed people had had a positive environmental influence on them, either through conversation and discussion or by being positive role models in their behaviour. It was mainly the interviewees' children that were mentioned here, as well as younger relatives and friends and acquaintances. Schoolchildren took home what they had learnt at school and nagged their parents to recycle and to choose environmentally friendly products. Sometimes, adult children were role-models for their parents. Some of the younger interviewees had been inspired to become environmentally committed or develop an eco-friendly everyday practice by their teachers. Environmentally active schoolmates were positive role-models in many cases, as were the friends who practiced organic cultivation.

Voluntary work and active involvement in social movements in general encouraged a handful of people to take an active interest in the environment and/or to move to an eco-village. These people had been active and had taken part in demonstrations or had other sorts of contact with the different alternative movements that emerged from the start of the 1960s. The ones mentioned most often were the anti-nuclear power campaign, and the peace and solidarity movements. Voluntary commitment and participation in social movements had, however, little effect on attitudes to the environment.

For a number of households, detailed information about eco-villages, contact with such villages or opportunities of moving to one were important and helped to raise their interest in the environment. The initial contact was not always the result of an expressed commitment to the environment, rather a first step towards an interest in eco-village life.

As for *experiences linked to transitions and turning points in a person's life course*, it was becoming a parent that seemed to have had the greatest impact on the everyday practice of the households under study. Parenthood, not surprisingly, gave rise to changes in the division of labour at home, the use of time and the organization of the household finances. It also prompted new values that affected everyday practice: some people chose to work less and renounce consumerism in order to have more time with their children. Many moved to a house or a flat near the woods or countryside mainly to avoid health risks, but also to give the children a chance to get to know about their natural environment. Having their first child made parents aware of different kinds of danger, particularly that relating to their child's health and the quality of food and the air. The desire to avoid such risks led to, or went hand in glove with, an urge to orientate the lives of the child and the family itself towards greater environmental friendliness. In one household, it was the arrival of the grandchildren that sparked the interviewees' environmental interest.

As for *public events and the socio-cultural circumstances of society in general*, the vast majority of interviewees said that they had been influenced by the general public debate on environmental problems. This included information and debate on environmental problems and their solutions, the environmental agenda of the different political parties, the prevailing environmental ideology etc. Nearly everyone admitted to having been worried by information and ideas about environmental dangers, and most of these said that this concern had made them try to change at least one component of their everyday practice in an environmental direction. This effect of the *zeitgeist* and media information corresponds closely with the results from various opinion polls showing that most people claim to have some sort of environmental interest. The prevailing proenvironmental ideology also makes such attitudes presumably socially desirable for the majority.

Only two households adopted a clearly sceptical stance on the problems and solutions dealt with by the media, their reason being that 'many environmental issues are symbolic'. A few interviewees gave no mention at all to any influence towards environmental friendliness from the *zeitgeist* or the media. However, no interviewee gave information, knowledge or *zeitgeist* as the only influence they had experienced. No interviewee singled out the *zeitgeist* as the decisive or only influence on them; rather, in most cases, it had reinforced environmental interest or allowed them to define their former way of making a living and everyday practice as environmentally friendly or unfriendly. Only a handful of people said that some feature of the *zeitgeist* had greatly affected their attitude towards the environment. On the other hand, about two thirds of the interviewees mentioned that their attitudes towards the environment were based on childhood experiences. There is of course a risk that the interviewees were reconstructing their own life histories to suit their current approach to life, and could have been focusing on elements of their life experiences that anticipated their current environmental position or that conformed with their ideas of how one should live in harmony with the environment.

In the interviewees' narratives about their life courses, a multitude of routes could be observed towards their current way of life and approach to the environment in their everyday practice. Changes in ways of life in general as well as environmental commitment and proenvironmental behaviour appear in a variety of ways. However, it is possible to distinguish several patterns in these life course changes: continuity, youthful renouncement followed by a certain amount of re-alignment with their childhood way of life, and a more lasting renouncement. Within the framework of these patterns of change, we have been able to distinguish a number of profiles on the basis of important experiences.

A high degree of continuity over the life course was characteristic of *older people who have always lived in a spirit of thrift*. This continuity applied mainly to keeping and making use of natural resources and thrift. Many of these people said that they had lived under three sets of historical conditions, with the austerity of their childhood years being replaced by the consumer society, itself succeeded by a phase of somewhat tougher and less predictable material conditions. These people became gradually, and relatively late on in their lives, aware of the environmental aspects of their own way of life, which had not changed very much despite the transitions from one set of conditions to the other. What had once been normal practice for these people, in later life had come to be seen as environmentally friendly. Continuity was also characteristic of *nature lovers*, who had a lasting relationship with nature as a common theme in everything they did. Experiences of 'nature' had been and still were important for these people and their interest in the environment was considerable, even though it did not necessarily lead to any environmental commitment, eco-village life or proenvironmental behaviour in their everyday practice. Comparatively small changes over the life course were also a feature of the *permanently indifferent*. Many of these people were dedicated professionals, while others spent much time and energy on working for societies and organizations. This life orientation had existed since early adulthood, and environmental issues were kept at a distance and viewed with a certain degree of cynicism.

In some cases, the life course change involved an early rejection of the lifestyle experienced when growing up and then its later, but in various ways, resumption. Following an exploratory period, these *reverters* re-adopted certain aspects of a local or regional way of life. Important elements of their lives as reverters included being close to nature, being able to feel familiar with the region and having local roots. The reverters had a certain degree of interest in nature, but the proenvironmental orientation in their everyday practice and expressed environmental commitment varied.

In two other groups, the life course change involved the persons dissociating themselves decisively and, for the time being at least, permanently from the circumstances that prevailed during childhood. One of these groups was the *student left*, in which ideological undercurrents had raised an awareness of environmental problems or other kinds of social issue, something which had had an impact on how these people had formed their way of life. These are people who were, or had been, radical activists. Explicit global solidarity was part of the pattern, as was neighbourhood co-operation; the importance given to the latter was also manifested in their everyday practice. Environmental commitment was high amongst representatives of the student left, and derived mainly from contact with or involvement in alternative movements. The proenvironmental orientation in their everyday practice was, however, not as noticeable. Permanent disassociation from childhood conditions was also a distinguishing feature of the *young thrifty persons*. Here, ways of life had been shaped by the immediate circle of friends and relatives. The small group representing this pattern was largely made up of younger people who had rejected the consumerism of their parents or their parents' generation. There were no radical political overtones, the transformation being mainly expressed in terms of 'leaving the rat race' and an interest in alternative values: close relations with children and other relatives, health and the environment. These changes are largely thought to be associated with starting a family and other life course transitions.

People's Proenvironmental Rationale

The embeddedness of proenvironmental action in everyday life contexts includes the incorporation of an environmental concern in everyday decision-making. In contemporary interpretations of environmentally-aware behaviour, it is assumed that individuals do not simply act in an environmentally friendly way when a situation provides material benefits: they place restrictions on their actions which dictate that the ecological alternative is chosen in preference to the financially beneficial, even when they are in mutual conflict (Merchant, 1992, pp. 61–81; Axelrod, 1994; Stern and Dietz, 1994; Buttel, 1987). We will now look at different ways in which environmental concern enters the decision-making process, our point of departure being an analysis by the English political scientist and philosopher Robert E. Goodin of how ethical principles can be incorporated into an individual's decision-making in general (Goodin, 1982, pp. 95–121).

If this is applied to the environment, consideration can take the form of a *prudential and enlightened self-interest*. People act ecologically because they expect to benefit from other people acting in the same way. However, it can also be *internalized*, and form part of an individual's utility function, becoming an element that is satisfied in the same way as any other. Acting in a manner consistent with an internalized ecological concern is its own reward. Finally, an ecological approach can be a *principle that is taken seriously*. When people take something extremely seriously, they generally differentiate it from a self-interested pursuit for material gains. They refuse to weigh up environmental friendliness against other behaviour from which they could derive gratification.

Our discussion of these different ways of expressing environmental consideration is based on the same in-depth interviews as in the previous section. One part of the interview was designed to reveal the way people reasoned about the factors that might influence their decision when faced with a dilemma in which proenvironmentalism competed with other considerations. A series of questions were asked concerning everyday dilemmas in which the environment is weighed up against other worthy considerations, and where, in other words, acting proenvironmentally involves some kind of sacrifice. Questions addressed their choices of washing up liquid in the supermarket, fresh food for the evening meal, or transport (car or bus) to work. In all cases the choice was between an ecological and a conventional alternative, whereby the former was more expensive or more time consuming. The questions asked how the interviewee would reason in such situations.

Enlightened self-interest does not appear to play any prominent role as a mechanism behind proenvironmental behaviour. The interviewees generally dismissed the idea that their own actions were dependent on enough people acting in the same way (Diekmann and Preisendörfer, 1992, pp. 245–246). As for the importance of what others do and think, however, there were significant variations in reasoning. At one end of the scale were people who stated very clearly that their decision was independently made. Their choice of environmental alternative did not depend on their thinking or knowing that other people were doing the same. They presented their behaviour as controlled by clear-cut rules that were completely unconditional on the actions of others. This approach is based on the idea that individual responsibility must be taken somewhere along the line, even though it may or may not be effective:

As far as I'm concerned, even if it doesn't make my local Konsum supermarket stock more organic vegetables, I have to take responsibility somewhere, even for my own sake. I don't want to buy anything else, and I'm a customer here, in which case they'll lose at least one little customer that wants organic and not the other stuff. And I think that the supermarket should make sure that they stock it. You've got to start somewhere, haven't you.

This attitude often coincides with a reluctance to try to influence others, believing that non-interference in other people's behaviour warrants non-interference in their own. In these cases, other people's behaviour is of very little consequence, if any. Proenvironmental action in everyday practice therefore appears as very much a private matter. It should be added that the emphasis on proenvironmental action as a private concern and one of personal choice can also be an expression for a desire to present an independent image. In the reasoning of other interviewees there was a degree of behavioural link with the actions of others, although not in any causal way; it was more an expression of the hope that others would do as they themselves did, their justification being that a combination of small actions could result in the achievement of a greater good. People realized, however, that their own contributions to a better environment are small or negligible. Yet, instead of breeding feelings of powerlessness in many people, this could, despite everything, stimulate a certain optimism: even though a personal contribution may be considered irrelevant it is still thought to have some significance, and actions which are believed to contribute to reducing the environmental load come across, therefore, as worth persisting with.

If proenvironmental actions are said to have only a weak relationship with the behaviour or thoughts of others, it is possible that they are performed for reasons of personal gratification, regardless of what others do. Internalization should, in turn, result in a pricking of the conscience when for one reason or another a person chooses a non-ecological alternative.

However, many people either denied that they would develop a bad conscience if they were to choose non-ecological alternatives, or said that they sometimes developed a bad conscience, or a slightly bad conscience, or said that they did so for certain things or in certain situations. The 'inner voice' did not seem particularly strong for the interviewees. Some said that they had a permanent bad conscience and that they were continually reminded to make the proper choice by the availability of ecological alternatives. A more common phenomenon was that the conscience operated selectively and was activated when faced with certain actions or in certain situations. The 'inner voice' seemed stronger for some people when faced with the choice of cleaning agent rather than food; for washing up liquid, the environmentally damaging effects were considered so tangible that the 'inner voice' of the conscience was aroused more often and more intensely than it was for food. Non-environmentally friendly washing-up liquid was associated with toxic substances released by the user into the environment (Solér, 1997).

A bad conscience seemed reasonably easy to bear for some people, and although there was remorse, it was transitory. Nor was the 'inner voice' of the conscience activated when on the odd occasion they acted against their better judgement. Single deviations from the right path were justifiable in terms of their minor significance in light of an otherwise general proenvironmental approach, leaving no cause to feel guilty.

> Basically, I think we make a conscious or subconscious attempt to think about acting in an ecological way, buying good quality goods and so on. I think that if there wasn't the stuff that you'd normally buy and you had to buy something that wasn't so environmentally friendly, I guess I'd think that as I do the right thing most of the time, this one time is just a drop in the ocean.

Some people answered that the link between their individual responsibility and their own consciences on the one hand, and solutions to environmental problems on the other had weakened with time. They had progressed from seeing environmental friendliness as an issue mainly of individual responsibility to placing more emphasis on circumstances beyond their control, believing that too much responsibility for the environment was placed on the individual. They saw our environment as sensitive to a multitude of impacting factors, factors that lay beyond the individual's sphere of responsibility. Developing a bad conscience from single events was described as a phase now passed, because it was apparently difficult for them to make their everyday practice work when the environmental consequences of their own behaviour constantly had to be borne in mind. Individual actions had to be seen in the right perspective. In such cases, the ethical principal of acting ecologically seemed to have lost its grip on the individuals in question, possibly because it had been weighed up against other beneficial aspects of their behaviour. There is an inherent danger here that the guiding power that principles have over behaviour will eventually weaken (Goodin, 1982, pp. 109–110).

It might be expected that a person who has internalized proenvironmental attitudes is prepared to compensate non-environmentally friendly acts in any one area with environmentally friendly ones in another, in order to reach the degree of environmental friendliness that he or she considers desirable. Intuitively, it seems reasonable to expect this when individuals are placed in situations in which different goals conflict with one another. Yet this is not what we find. For many of the interviewed, compensation activity was manifestly almost akin to cheating; you should act proenvironmentally, regardless of whether your 'eco-account' is in the red or the black:

No, not really because you have to do it anyway. No, it's not about compensation, you can cheat sometimes, but you can't compensate. You can't. No, no, I don't think you can say to yourself that I'm now doing this because I did that, because I'd have done this whatever.

It seems that acting proenvironmentally was considered more of a duty, and nobody should be able to buy their way out of having to act proenvironmentally in one area by compensatory actions in another. For these people, each action is a discrete entity that cannot be compared with or offset against another. The reasoning in some cases was that each proenvironmental action, however small it might appear, was of value, especially when combined with others. Some people argued that certain actions were so important that they could never be compensated for: proenvironmental actions came naturally to the environmentally aware and were an integral part of a person's identity as an ecologically sound individual. In this can be detected the idea that consistency in sticking to the environmentally friendly alternative is essential in order to avert the danger of it becoming devalued in importance through comparison with other behaviours that offer gratification. Once the decision had been made to take the ecological alternative in one particular area, it had to be stuck to.

The tendency to let certain behaviour be strictly governed by proenvironmentalism, in that environmental concern is taken particularly seriously, was found in a little less than half of the interviewees and never extended to anything other than only limited areas of the everyday practice. These areas included a refusal to buy non eco-labelled cleaning agents, disposable bottles, leaded petrol, cans, products packed in certain types of packaging and bleached toilet paper or to travel by car to work. As one can see, these are all aspects of people's everyday practice for which, at least in some cases, there has been a prolonged and occasionally intense public debate on environmental problems.

Our analysis of people's views of how proenvironmentalism enters their everyday decisions suggests that, in general, self-interest determined by the similar actions of others is apparently of minor importance. The same can be said for internalized proenvironmental principles; and according to the interviewees, the voice of the conscience seems to operate in a selective and largely transitory way.

What seems to play a greater role is the third approach to caring for the environment – the singling out of certain areas or actions as targets for environmental friendliness. The interviewees stressed the necessity of remaining loyal to the proenvironmental actions that had been decided upon and of not renegading on this behavioural policy. Doing so gives each action, no matter how small, a value disproportionate to the direct impact it has on the environment. This probably concerns actions that people have managed to pick up but which have not yet become established routines (Læssøe, Hansen and Søgaard Jørgensen, 1995, pp. 41–44; Biel, 1999).

In the modest sample we were working with there were no marked differences in the ways of reasoning about how environmental concern enter their decision-making between people with differing degrees of environmental commitment. Nor was there anything special about the way most of the interviewees living in eco-villages reasoned about the proenvironmentalism in their everyday practice. Moreover, we could not detect any significant difference between women and men in their way of reasoning, despite expectations to the contrary given that the women had a much greater effect on the households' environmental orientation. It is worth noting, however, that the women were often more capable than the men of expressing their reasoning on environmental issues.

Conclusions

We have been discussing the importance of different kinds of everyday life context as suggested by an extensive survey of households in Stockholm; we have looked in part at the broader context that household cultures represent and in part at the more narrow contexts described as life course phase and gender order. We have argued that variations within the life course phase and gender categories correlate partly with differences in household culture, and have found these ideas supported by the results of the survey. The total explanatory power of the contexts we studied is, at its highest, just over 20 per cent in the models we have presented. The fact that it is no higher can partly be due to the small number of indicators used to categorize household cultures, but it still suggests that five household cultures is not enough to encapsulate the diversity of their everyday practice. It is possible that the different components united by the theory do not work without the presence of other combinations of, for example, practice and meaning. The results do suggest, however, a certain 'order in the chaos'.

The second study of a limited number of households used in our discussion suggests that life course experiences are important in determining a person's environmental beliefs and attitudes, albeit in different ways and with differing results. The route to proenvironmental orientation in material conditions and everyday practice takes more the form of a series of minor steps than of one major choice. People who stated that their ecological path began during childhood generally said that it was interwoven with other orientations in everyday practice. Environmental friendliness as a way of life only became obvious later, with the impacting factors coming from indirect sources. Many of these people could trace a continuous route towards the proenvironmental nature of their everyday practice, from childhood to the present. Some people saw the emergence of proenvironmental orientation in everyday practice as something that became more entrenched as their environmental awareness developed. Others explained that childhood experiences of unintended proenvironmentalism, at one time forgotten or rejected, had been reawakened at a later stage of their lives. Differences in the early history of proenvironmental orientation suggest that for some it is a matter of personally stored memories and experiences, for others that of a conscious and considered processing of experiences associated with changes in the world around them.

Our discussion of household culture and the different ways of reasoning around proenvironmentalism is based on two separate surveys, which makes any direct link between these two types of everyday life context difficult to make. Yet it might be possible to make some tentative suggestions. Both hierarchical and egalitarian households can be expected to take the environment particularly seriously by allotting certain clearly-demarcated areas to their proenvironmentalism, albeit for different reasons and to different extents. For the hierarchist, these areas are defined and limited by the opinions of experts and authorities. The egalitarian, on the other hand, has no such restrictions and is likely to possess a much wider proenvironmental scope, lacking as he or she does any externally established criteria for defining the areas for proenvironmental action. In their day to day lives, egalitarians are naturally forced to weigh up the demands of their proenvironmentalism against those of other goals. In the extreme egalitarian case, with close scrutiny paid to everything that enters and leaves the household, every action can become a dilemma.

Individualists can hardly be expected to reason in the same way as hierarchists and egalitarians. Enlightened self-interest is the likely force shaping the individualist's way of thinking. An individualist acts proenvironmentally if enough other people do so as well. Individualistic household cultures can, however, also be expected to internalize proenvironmentalism and weigh it up against other interests, since there is an element of calculation behind their behaviour. Fatalistic household cultures, which never really have to face any such dilemmas, simply take the alternative that happens to present itself. Hermits probably vacillate from one way of reasoning to another, and can chiefly be expected to internalize proenvironmentalism and even take the environment seriously, similarly to egalitarians but without the latter's social goals and resolute attitudes.

It is worth mentioning that people's proenvironmental reasoning displays elements reflecting an expression of concern that principles might weaken over time; that they would be unable, in other words, to maintain their proenvironmental behaviour in selected areas. They seem worried that they might drop behind and lose ground, not least because environmental considerations often have to be weighted up against those of other objectives, and that such comparisons threaten to undermine proenvironmentalism. This puts individuals in a situation for which they stress that excessive responsibility is placed on them. In the Stockholm survey, many people suggested that households had done their bit and that the authorities and business should take responsibility for the environmental adaptation of urban transport and other such matter of socio-environmental importance.

Finally, we would like to mention some consequences for sustainable development. The importance of different kinds of everyday life contexts for environmental orientation indicates that it is not an easy task to bring about substantial changes in everyday practice. What households are willing to do in the interest of the environment depends critically on life course experiences, current life course phase and physical infrastructure.

Any plan of action towards sustainable development should be designed on the grounds that there are vast differences between people as regards the importance of environmental concern in their everyday practice and meaning. People in different household cultures will contribute to such a development in certain ways and will be prepared to shoulder responsibilities of different kinds. Diversity is important in another sense too. If sustainable development is to be promoted, reasons other than environmental friendliness must be deployed; for even if people want to be good citizens in the environmental field, more is needed to create a stable foundation upon which to build. Compelling combinations of motives must be sought, one such that has proved effective being the environment and health. Building upon the situations offered by different life course phases is also effective, such as the concern felt by parents for the health of their babies and the activism of youth.

The observable weakening of proenvironmentalism with time can be slowed inasmuch as it is supported by other motives. However, it can also be partly substituted by solutions which lift the burden of responsibility from the individual or household by integrating environmental adaptation into the infrastructure.

The importance of gender order is problematic. Certain aspects of environmental policy result in households becoming overburdened with tasks, particularly the women. This is effective if one believes that women will accept, on the basis of their 'caring' way of thinking, extra work on top of the unpaid housework they already do. The problem here is that women can tire of sorting waste etc. It has to be remembered that many of the measures that are proposed are already part of the prevailing gender order. Besides it being wrong (in our opinion) to exploit such an order, aggravating an already tense situation can create problems for environmental policy. If planning continues to operate within the framework of the established order, it would be a more pressing task for environmental policy to formulate solutions that engage men more, by appealing, for instance, to their interest in technology.

References

Astrom, L., (1986), *I kvinnoled. Om kvinnors liv i tre generationer* (Women's life in three generations), Malmö.

Axelrod, L. J., (1994), 'Balancing Personal Needs with Environmental Preservation: Identifying the Values that Guide Decisions in Ecological Dilemmas', *Journal of Social Issues*, vol 50.

Back-Wiklund, M., (no year specified), 'Det moderna lokalsamhället. Dokument over en svunnen tid med sikte pa det moderna' (The modern local community. A study on times past with an eye to modern society), in *Arbetets organisering, manniskans forsorjning. Vanbok till Bengt Rundblad* (Work organization and livelihood), Göteborg.

Beardsworth, A., and Keil, T., (1992), 'The vegetarian option: varieties, conversions, motives and careers', *The Sociological Review*, vol 40.

Bennulf, M., (1994), *Miljoopinionen i Sverige* (Environmental opinions in Sweden), Lund.

Berge, G., (1998), 'På biltur med Weber. Bilkjøring som sosial handling' (On the road with Weber. Car driving as a social act), *Sosiologi i dag*, vol 28.

Berger, I. E., (1997), 'The demographics of recycling and the structure of environment behavior', *Environment and Behavior*, vol 29.

Biel, A., (1999), 'Do people choose to be environmentally friendly?' in L. J. Lundgren (ed.), *Changing environmental behaviour*, Stockholm.

Bjerén, G., (1981), 'Female and male in a Swedish forest region: old roles under new conditions', *Antropologiska studier*, no. 30/31.

Bjerén, G., (1987), 'Kvinna och man, natur och kultur i ett svenskt småbrukssamhälle' (Female and male, nature and culture in a Swedish local community), in D. Kulick (ed.), *Från kon till genus. Kvinnligt och manligt i ett kulturellt perspektiv* (From sex to gender. Female and male in a cultural perspective), Stockholm.

Bjerén, G., (1991), 'Livsformer och samhallsforandring i Sverige' (Life-modes and social change in Sweden), *Kvinnovetenskaplig Tidskrift*, arg. 12.

Bourdieu, P., (1979), *La distinction. Critique sociale du jugement*, Paris.

Bourdieu, P., (1980a), *Le sens pratique*, Paris.

Bourdieu, P., (1980b), 'L'identité et la représentation. Eléments pour une réflexion critique sur l'idée de région', *Actes de la recherche en sciences sociales*, vol 35.

Brand, K.-W, (1997), 'Environmental consciousness and behaviour: the greening of lifestyles', in M. Redclift and G. Woodgate (eds), *The International Handbook of Environmental Sociology*, Cheltenham, UK, Northampton, MA, USA.

Buckingham-Hatfield, S., (2000), *Gender and Environment*, London, New York.

Buttel, F. H., (1987), 'New Directions in Environmental Sociology', *Annual Review of Sociology*, vol 13.

Connell, R. W., (1987), *Gender and power*, Cambridge.

Connell, R. W., (1990), 'A whole new world: Remaking masculinity in the context of the environmental movement', *Gender & Society*, vol 4.

Dahlgren, L., and Lindgren, G., (1987), *De sociala gransridarna – om konverterings- och anpassningsstrategier i ett lokalsamhälle i Tornedalen* (Actors in the social borderland – strategies of conversion and adjustment in a local community), Umea.

Dake, K., and Thompson, M., (1993), 'The meanings of sustainable development: Household strategies for managing needs and resources', in S. Wright *Human Ecology: Crossing Boundaries*, Fort Collins.

Dake, K., Thompson, M., and Neff, K., (1994), *Household Cultures*, Household Cultures Project, Berkeley (manuscript).

De Vault, M. L., (1991), *Feeding the Family: The Social Organisation of Caring as Gendered Work*, Chicago and London.

Derksen, L., and Gartrell, J., (1993), 'The social context of recycling', *American Sociological Review*, vol 58.

Diekmann, A., and Preisendorfer, P., (1992), 'Personliches Umweltverhalten. Diskrepanzen zwischen Anspruch und Wirklichkeit', *Kolner Zeitschrift für Soziologie und Sozialpsychologie*, vol 44.

Dietz, T., Stern, P. C., and Guagnano, G. A., (1998), 'Social Structural and Social Psychological Bases of Environmental Concern', *Environment and Behavior*, vol 30.

Douglas, M., (1982), 'Cultural Bias', *In the Active Voice*, London.

Douglas, M., Gasper, D., Ney, S., and Thompson, M., (1998), 'Human needs and wants', in S. Rayner, E. L, Malone (eds), *Human choice and climate change. Volume one. The societal framework*, Columbus, Ohio.

Durning, A. T., (1992), *How much is enough? The consumer society and the future of the earth*, London.

Ehn, B., (1992), 'Livet som intervjukonstruktion' (The construction of life in interviews), in Tigerstedt et al. (eds), *Självbiografi, kultur, liv. Levnadshistoriska studier inom human- och samhallsvetenskap* (Autobiography, culture, life. Studies in life history in the humanities and the social sciences), Stockholm.

Ellis, R. J., and Thompson, F., (1997), 'Culture and the Environment in the Pacific Northwest', *American Political Science Review*, vol 91.

Elvin-Nowak, Y., (1999), *Accompanied by guilt. Modern motherhood the Swedish way*, Stockholm.

Finger, M., (1994), 'From Knowledge to Action? Exploring the Relationships Between Environmental Experiences, Learning, and Behavior', *Journal of Social Issues*, vol 50.

Förpackningsinsamlingen, SIFO (the Swedish Institute of Public Opinion Research), (1999), *Fragor till allmanheten om forpackningsatervinning och atervinningsstationer* (Survey on recycling), SIFO, Stockholm (stencil).

Frønes, I., et al., (1997), *Livsløp. Oppvekst, generasjon og social endring*, Oslo.

Giele, J. Z., and Elder, G. H., (eds), (1998), *Methods of Life Course Research*, Thousand Oaks.

Gilligan, C., (1982), *In a Different Voice: Psychological Theory and Women's Development*, Cambridge, MA.

Godbey, G., Lifset, R., and Robinson, J., (1998), 'No Time to Waste: An Exploration of Time Use, Attitudes toward Time, and the Generation of Municipal Solid Waste', *Social Research*, vol 65.

Goodin, R. E., (1982), *Political Theory and Public Policy*, Chicago.

Grendstad, G., (2001), 'Nordic Cultural Baselines – Accounting for Domestic Convergence and Foreign Policy Divergence', *Journal of Comparative Policy Analysis*, vol 3.

Haavind, H., (1982), 'Makt og kjaerlighet i ekteskapet', in *Kvinneforskning: Bidrag til samfunnsteori*, Oslo.

Haavind, H., (1984), 'Fordeling av omsorgsfunksjoner i småbarnsfamilier', in *Myk start – hard landing. Om forvaltning av kjønnsidentitet i en endrigsprocess*, Oslo.

Haavind, H., (1992), 'Vi maste söka efter konets forandrade betydelse' (We must look for the changing importance of gender), *Kvinnovetenskaplig Tidskrift*, arg 13.

Hägg, (1993), *Kvinnor och män i Kiruna. Om kon och vardag i forandring i ett modernt gruvsamhälle 1900–1990* (Women and men in Kiruna. On changes in gender and everyday life in a modern mining community 1900–1990), Umea.

Hagman, O., (2000), *Bilen, naturen och det moderna. Om natursynens omvandlingar i det svenska samhallet* (Car, nature and modernity. Transformations of views on nature in Sweden), Goteborg.

Hallin, P.-O., (1994), *Grona konsumenter och nygamla utopister. Hushall och miljo i Sverige och USA* (Green consumers and neotraditional utopians. Household and environment in Sweden and the USA), Lund.

Hjorthol, R., (1998), *Hverdaglslivets reiser. En analyse av kvinners og menns daglige reiser i Oslo*, Oslo.

Hjorthol, R., (2001), 'Gendered Aspects of Time Related to Everyday Journeys', *Acta Sociologica*, vol 44.

Højrup,T., (1983a), *Det glemte folk,* Hørsholm.

Højrup,T., (1983b), 'The Concept of Life-mode. A From-Specifying Mode of Analysis Applied to Contemporary Western Europé', *Ethnologia Scandinavica*, vol 22.

Holmberg H.-E., (1999), *Rapport: Konsumentundersökning om ekologiska produkter/ KRAV* (A survey on the consumption of organic food), LUI Marknadsinformation Lund (stencil).

Juntti-Henriksson, A.-K., (1996), *Kvinnor, kultur och framtid i Tornedalen* (Women, culture and the future of Tornedalen), Kvinnokraft, Overtorneå.

Kalof, Dietz, T., Stern, P. C. and Guagnano, G. A., (1999), 'Social Psychological and Structural Influences on Vegetarian Beliefs', *Rural Sociology*, vol 64.

Kempton; W., Boster, J. S., and Hatley, J. A., 1995, *Environmental Values in American Culture*, Cambridge, Mass, London.

Læssøe, J., Hansen, F., and Søgaard Jørgensen, M., (1995), *Grønne Familier. Miljøvenlige levemader – og mulighederne for at støtte udviklingen af dem,* Danmarks Tekniske Universitet.

Lindén, A.-L., (1994), *Människa och miljö. Om attityder, värderingar, livsstil och livsform* (Man and environment. On attitudes, values, lifestyle and life-mode) Stockholm.

Lindén, A.-L., (1996), 'Fran ord till handling. Individuella mojligheter och samhalleliga restriktioner' (From words to deeds. Individual opportunities and social constraints), in L. J. Lundgren (ed.), *Livsstil och miljo. Fraga, forska, forandra* (Lifestyle and environment. Asking, researching, changing), Stockholm.

Lunt, P. K., and Livingstone, S. M., (1992), *Mass Consumption and Personal Identity. Everyday Economic Experience*, Buckingham, Philadelphia.

Mannheim, K., (1972 [1952]), *Essays on the Sociology of Knowledge*, London.

Mars, G., and Mars, V., (no date), 'The creation of Household Cultures', reported in Douglas, M, 1996, *Thought styles*, London, Thousand Oaks, New Delhi.

Martensson, M., and Pettersson, R., (2002), *Gron vardag. Hushall, miljohansyn och vardagspraktik* (Greening everyday life. Households, environmental concern and everyday practice) Stockholm.

Mellor, M., (1997), *Feminism and Ecology*, Cambridge.

Merchant, C., (1992), *Radical Ecology. The Search for a Liveable World*, New York.

O'Rand, A. M., and Krecker, M. L., (1990), 'Concepts of the Life Cycle: Their History, Meanings, and Uses in the Social Sciences', *Annual Review of Sociology*, vol 16.

Osterman, T., (1998), *Opinionens mekanismer. Om varderingar och verklighet* (On the mechanisms of opinion), Stockholm.

Pahl, R. E., (1984), *Divisions of labour*, Oxford.

Pedersen, K., (1993), 'Gender, Nature and Technology: Changing Trends in "Wilderness Life" in Northern Norway', in R. Riewe and J. Oakes (eds), *Human Ecology: Issues in the North*, Edmonton.

Pedersen, K., (1995), 'Pa sporet av et mangfold friluftslivsstiler', in Damkjer, S og Ottesen, L, *Ud i det fri. Sport, friluftsliv, turisme*, Odense.

Plumwood, V., (1993), *Feminism and the Mastery of Nature*, London.

Roos, J. P., 'Livet – berättelsen – samhallet: en bermudatriangel?' (Life – narrative – society: a Bermuda triangle), in Tigerstedt et al. (eds), *Sjalvbiografi* (Autobiography).

Rydenstam, K.,(1992), *I tid och otid. En undersokning om kvinnors och mans tidsanvändning 1990/91* (A survey of time use of women and men 1990/91), Stockholm.

Sandberg, A., (1998), *Moralutveckling och gender* (Moral development and gender), Sociologiska institutionen, Stockholms Universitet, Stockholm.

Solér, C., (1997), *Att kopa miljovanliga dagligvaror* (On buying ecologically friendly consumer goods), Stockholm.

Sorbom, A., (1999), 'Engagemang med förhinder. Yngre och aldre om politik, forandring och miljo' (Commitment contained. Young and old people on politics, social change and the environment), in *Det unga folkstyret, SOU 1999:93* (Young people and democracy), Stockholm.

Steg, L., and Sievers, I., (2000), 'Cultural Theory and Individual Perceptions of Environmental Risks', *Environment and Behavior*, vol 32.

Stern, P. C., and Dietz, T., (1994), 'The Value Basis of Environmental Concern', *Journal of Social Issues*, vol 50.

Svenskarnas resor (Travels in Sweden), (2000), Statistiska Centralbyran, Stockholm.

Thompson, M., Ellis, R., and Wildavsky, A., (1990), *Cultural Theory*. Boulder: Westview Press.

Thurén, B.-M., (1996), 'Om styrka, räckvidd och hierarki, samt andra genusteoretiska begrepp' (On force, scope, hierarchy and other theoretical gender concepts), *Kvinnovetenskaplig Tidskrift*, arg. 17.

Thurén, B.-M., (2000), 'On Force, Scope, Hierarchy. Concepts and questions for a cross-cultural theorization of gender. Paper presented at the 4th European feminist conference, Bologna 2000', Umeå University.

Tigerstedt et al. (ed.), (1992), *Sjalvbiografi, kultur, liv. Levnadshistoriska studier inom human- och samhällsvetenskap* (Autobiography, culture, life. Studies in life history in the humanities and the social sciences), Stockholm.

Waara, P., (1996), *Ungdomar i gransland* (Youth in a border community), Umeå.

Warde, A., (1997), *Consumption, Food and Taste. Culinary Antinomies and Commodity Culture*, London, Thousand Oaks, New Delhi.

The Formation of Green Identities – Consumers and Providers

Anna-Lisa Lindén and Mikael Klintman

Introduction

In many fields of consumption the consumer has a first choice of either purchasing or avoiding products. But there are important exceptions from this freedom of choice. In the utility sectors almost every supply of goods and services includes necessary consumption, at least to a certain extent. As a member of society you have to get rid of your refuse in a safe way by using the existing supply of waste services. To a certain degree every citizen in modern society has to use electricity and water. Consumption of utility services and goods represents an enormous market involving a very large number of consumers. Greening the utility sectors thus ought to play an extremely important role in strategies of greening society.

Diversity of choice is, broadly speaking, nothing new – at least not regarding waste. Electricity, being a fruit of modernity, has from its beginning been provided as one product, leaving out possibilities of choice. Waste management reveals a comparable history. Pre-modern and rural communities composted the organic waste and used it as fertilizers on the field. Much of the waste from food was given to the pigs; metal and glass were frequently reused (Rosén, 1988). Such separation was institutionalized. However, in Sweden for example, the separation more or less ended in the 1920s when artificial materials and synthetic fertilizers started to be used. The synthetic materials were too difficult for farmers to reuse (Johansson, 1997:196). Aside from particular projects of newspaper and glass recycling, convergence into one waste 'fraction' has been a general pattern of the twentieth century (DOMUS, 1998).

In this article we explore the late modern tendency of differentiating electricity sources and waste fractions. It would be unfair to describe contemporary processes of differentiation as regressions to pre-modern vague fragmentation. Instead, current differentiation of products and tariffs involve the ambition of standardized choices founded on reflexive thinking of consequences for health, environment, and economic profit.

Product and tariff differentiation (PTD) refers to cases in which a previously one-fold utility supply has become diversified. We examine the supposition that the one-sided model of 'offering new choices to consumers' has lost its significance in the two utility sectors. It implies that co-creative provisions of choices are being developed in the utility sectors, provisions involving both providers and consumers. Other specifically environmental assumptions are closely related to product and tariff differentiation. One assumption is that new product and tariff choices lead consumers to increasing environmental awareness.

The suggestion that product and tariff differentiation (PTD) would be a powerful part of ecological modernization is based on the following reasoning. As products and tariffs in the utility sectors become differentiated, the public is confronted with new choices of products and services. These new choices might make the public more active. The question is what the possibilities and impediments are to an active involvement in practical contexts. This regards both technical involvement and new types of political interaction between citizen-consumers and providers. In the aggregate, the issue of product and tariff differentiation (PTD) as consumer empowering raises the question of whether or not such new types of involvement have the potential of bringing about environmentally beneficial system change. It is important to note that evaluations of innovations have many faces, founded on perceptions by various actors and interests. Moreover, it turns out to be perceptive to tendencies that in certain cases run parallel with differentiation, namely convergence and standardization of choices.

Ecological modernization theories mostly deal with macro level questions, for example how greening politics introduce new policies in enterprises and a supply of products and services which take environmental aspects into consideration. The micro level questions deal with questions about how individuals prepare themselves to be environmentally aware of green consumers asking for, buying and using green products. The process includes consumers as well as providers in relations, which include ecological, environmental, economic and social aspects. The main focus of the analysis in this article is on the micro level analysis in formation of green identities among consumers and providers in the waste and electricity sectors. Product and tariff differentiation have the function of catalyst in this process by adding green, economic and social values to products with a history of very low differentiation and possibilities of alternative choices.

The analysis starts with the identity perspectives, followed by an analysis of materialization of products in waste and electricity sectors. In the third and fourth section the relations are examined between the co-creation of green identities and behaviours of providers as well as consumers in greening processes within the utility sectors.

Product and Tariff Differentiation: A Green Identity Perspective

Product and tariff differentiation are closely related to phenomena of interest to various disciplines in the social and behavioural sciences. The Economic Man outlook emphasizes the significance of economic rationality in people's choices as products and tariffs become diversified. Behaviourism and cognitive social psychology focus on incentives and disincentives for greener consumer behaviour on a differentiated market. Within behaviourist environmental research it is standard to use terms like *behaviour change techniques*, and *behavioural intervention*.[1] One facet this type of research may have in common with our position is an interest in the concrete and manifest cause of environmental deterioration: human behaviour. The fact that human action is directly visible allows a purely behaviour-oriented researcher to avoid some of the methodological difficulties that, for example, studies of environmental attitudes can raise.

A philosophical objection to behaviourism frequently refers to its claim that the specific, external situation is the absolute determinant of all behaviours. Behaviourism is hence hardly compatible with a scientific interest in human freedom of choice and creativity (Joas, 1996:2).[2] The cognitive school, however, takes an interest in inner, creative processes and formations of meaning within an individual. Accordingly, the concept of *behaviour change techniques* has been replaced by *motivational techniques*.[3] Sayer (1979) sheds light on the separation of *behaviour* and *action*.

> By 'behaviour' we mean nothing more than a purely physical movement or change, such as falling asleep, breathing, that is, doing things which lack 'intrinsic' meaning structure. In contrast, doing which we call 'actions' are not wholly reducible to physical behaviour even though they may be coupled with it. Actions are constituted by *inter-subjective meanings*: putting a cross on a ballot paper, conducting a seminar, getting married, arguing, doing arithmetic, going on a demonstration are all examples of doings which nature *depends* on the existence of certain inter-subjective meanings.
> (Sayer, 1979:20–21.)

The cognitive perspective's focus on attitudes will be especially useful to study when they are combined with other variables. The question is *what* combination is the most efficient one. Wachs (1991:336) has suggested the technique of 'market segmentation'. People are grouped according to demographic and socio-economic characteristics in combination with their local circumstances (e.g., local policy, and

---

[1]See De Young (1993). See also Dwyer and Leeming (1993).
[2]The very title of Skinner's (1971/1978) best-selling book in behaviourism, *Beyond freedom and Dignity*, is sufficient to indicate a behaviourist viewpoint of human practices.
[3]See e.g. Geller, Winett and Everett (1982).

physical structure). These variables together are held to explain a significant part of the relationship between attitudes, motives and lifestyle choices.[4] Policy decisions can radically change people's motives for and against adopting more ecologically sound practices. To get a picture of motives for and against practice is especially important for policy makers.

Although the approach in this study shares the interest in human behaviour as well as the interest in human creative processes, its foci differ in fundamental ways from the other schools. A crucial difference is that these schools are generally founded on a top-down perspective. Accordingly, they may generate queries such as: 'How can leading institutions of society make people change their everyday habits in an environmentally respectful manner?' (Klintman, 1996:9). In other words, studies in the traditions mentioned mainly examine which initiatives by policy makers, local authorities or companies, are accepted by consumers.

The questions are broader in this study, for instance: 'How can society improve the conditions for its members to actively participate in environmental improvements? How can organisations of society be modified to motivate citizens taking their own environmental initiatives, initiatives that sometimes go further and maybe in other directions than governments appreciate?'[5] We aim at illuminating the relations between the levels of provider organizations and on the other hand individual consumers, that is how product and tariff differentiation (PTD) is co-produced at more than one actor level. It is not merely a matter of top-down versus bottom-up, but rather of how the differentiation process consists of a complex of linked green identities and images among consumers and providers.

While the attitudes of actors represent their views about a phenomena, identity represents the actors', providers' or consumers', internal evaluation of own attitudes and behaviours as green. On the other hand the other part in the relation, either he is a consumer or provider, may define the counterpart as having a green image. Thus green image represents the consumer evaluation of the provider identity or the provider evaluation of consumer identity. Their evaluation of activities, either it is supply of services and goods or choice of consumption, is thus closely related to each other.

Utility sectors differ from other sectors in their regulatory bases. Yet it would be a serious over-simplification to hold that it is merely the *comprehensiveness* of

---

[4]Based on Dobson and Tischer (1976), and Dobson and Nicolaidis (1974).

[5]See Joas (1996) for a comprehensive analysis of the concept of *action* in sociology. Joas holds that the sociological interest in human action has since Comte been an attempt *to 'limit the legitimisation of the principle of "laissez-faire" in the vulgarised forms in which classical economics has permeated European thought'* (p. 36). Comte aimed at bringing forth a normative and moral dimension of action, hoping to moderate the dominant perspective of 'rational' (as individualist and solely economic) action.

regulation that separates the two. Instead, utility sectors have traditionally been subject to *different* regulatory frameworks, commonly founded on ideals of free competition, health, safety and environmental concerns. These regulations aim both at extending and limiting the choices of consumers.

The questions raised can be summarized as follows:

- How closely is product and tariff differentiation (PTD) connected to liberalization in the two sectors?
- What groups of customers are targeted by the providers and by the customers already involved?
- What are the principal motives for and against introducing and choosing green products/tariffs?
- What different kinds of consumer identities and provider images related to product and tariff differentiation (PTD) can be distinguished, and how are they tied to the two sectors?

Co-Creation of Green Identities

Policy Background: Conflicting Views of Green Responsibility in the Utility Sectors

A general issue in society, which overlaps several of our themes, is illustrated by the following question: Should consumers or providers have the main responsibility of ecological sustainability in the two sectors? The strong-state ideal holds that legal restrictions and regulations on providers would do the better job. In contrast, the market economy ideal maintains that consumers ought to have free consumption choices and thereby the ultimate environmental responsibility.

When asking our interview persons[6] about who should have the main responsibility for environmental consequences of household consumption, the interviews reveal a hybrid position. Consumers both maintain that state and providers, including the consumers themselves, ought to have the main responsibility. At the same time, everyone is aware of the fact that a significant percentage of the public still chooses environmentally unsound products. Would it be fair to say that such product choices are signs of consumers' favouring environmentally sound and unsound supply? Or could it be interpreted as the very opposite, consumers giving a hint of what happens in a system of amoral market differentiation? Regardless, we will give examples of the crucial roles of combining provider and consumer responsibility in providing green tariff and product choices.

[6]See *Note* in the end of the text (p. 88)!

Is Green Product and Tariff Differentiation (PTD) Dependent on Liberalization?

When investigating ecological modernization and the utilities it becomes relevant to ask how closely product and tariff differentiation is connected to liberalization in the sectors of electricity, and waste. By presenting our cases we point at the fact that product and tariff differentiation (PTD) doe; not presuppose a liberalized market.

Liberalization in the *electricity sector* is common in the EU countries. The Swedish deregulation happened in 1997. Re-regulation, at least in the Swedish energy companies, has meant that the companies have developed their environmental improvements gradually. A reason might be, they believe at the Swedish Energy Company Electra, the new competitive situation. Re-regulation has changed the conditions for the energy companies so that, among other factors, the environmental identity has gained importance. For instance, the Swedish Energy Company Electra currently works actively with environmental certification according to ISO 14,000. The definition of green electricity varies between countries. Thus water-produced electricity would be considered as green electricity. In Sweden, however, it is not considered as green. Solar energy and wind power produced energy are considered to be green products. Both represent small fractions of produced energy. In our case wind power produced energy represents greening of the electricity sector.

As to *wind power* in Sweden, the preparation at the Electra Company started 1986–1987 long before the liberalization of the electricity sector. The administrative preparations took 3–4 years. In 1990, the first two windmills were constructed: Annica and Beatrice. They were among the largest 'smaller' ones. It was the first project constructing more than one turbine in the same area.[7] The local government took the initiative, when the company was part of the municipality. The authority's demand was five windmills should be established generating in total 1.5 MW. The company realized that they would not be able to establish and run five windmills with the profit requirements that they had set. This is why they initiated a wind power co-operative in the area. The company initiated and was active until the co-operative board was established. The Swedish Energy Company Electra symbolically still owns one of the 900 shares in the co-operative.[8] This

---

[7]Subsequently, they fulfilled the agreement with the local authorities in Lund, by building a third windmill. Moreover, the Swedish Energy Company Electra was active in helping a wind power co-operation to build their windmill, which makes four windmills.

[8]In addition to wind power, they say that they also produce other kinds of green electricity, since hydropower from plants constructed before 1996 are defined as green. They have done this all along, probably regardless of the environmental aspects of hydropower. Yet they believe in differentiating the products on basis of what it actually is rather than place the production sources under definitions such as green electricity. It is better that the consumers interpret whether or not the different sources are green or not. From one of their competitors it is possible to buy distinct nuclear produced power.

example shows that product differentiation is not necessarily connected with liberalization.

The Swedish waste organization studied illustrates how product and tariff differentiation (PTD) in the *waste sector* does not presuppose a liberalized market. The waste sector was deregulated in 1994. However, the municipalities still have the ultimate responsibility for waste collection. It might nevertheless be true that the local authorities are inspired by new options emerging in re-regulated utility sectors. As of today, reductions of the waste collection fee have got more prevalent in municipalities around the country. How common it is that households ask for this alternative remains to be examined, if it is approaching 'normalization'. It is without a doubt part of the trend towards a more differentiated waste policy, yet not followed by a re-regulated waste collection market from the viewpoint of the households. Although the municipalities have contractors doing the waste collection, the municipalities nevertheless have the final responsibility of getting it done in an organized and sanitary manner (according to the Waste Collection and Disposal Act, 1979:596; Section 2a, 1990:235). In sum, product and tariff differentiation (PTD) does not presuppose a liberalized market. However, competition tends to stimulate consumers and providers to introduce new choices on open utility markets.

Identities and Practices of Choosing Green Products and Tariffs

The Roles of Identity

To discuss identities in relation to consumption choices has become commonplace in cultural studies and sociology. However, because a vast majority of such studies have focused on clearly cultural goods, for example leisure activities, clothing, music preferences, it would be too bold to infer similar identity processes for all consumption from spheres, which so obviously are tied to identity building. Alan Warde stresses the importance of not underestimating the routine character of certain consumption:

> Although some people may attempt to create total life-styles as expressions of personal identity, most, despite the intention of advertising agencies, probably see choices between soaps or soups as not seriously prejudicial to their self-image.
> (Warde, 1992:25)

This true, we still maintain that the main problem is not that identity is assumed to lie behind all consumption, but rather that the concept of identity is often used carelessly. As Campbell (1995) notes, several confusions are involved when identities and consumption are explored. Firstly, the intelligibility of consumption is often wrongly assumed by necessity to imply a consensus as to what a certain consumption pattern means. Secondly, it is a mistake to compare messages through consumption.

Finally, the fact that a subject observes a message does not have to mean that the other subject get the intention to act or send a message (Campbell, 1995).

As big a mistake as it is to ascribe all consumption to social signs and messages, it is equally fallacious to reject the role of identity altogether within the utility sectors, merely based on the intuition that many consumer activities appear to be quite routine-based. The formation of a green identity and self-evaluation by a consumer as well as by a provider may be consciously done. New consumer research accordingly gives the identity concept new nuance to include more than status signs and symbolic utility. At present, consumption and ownership are increasingly dealt with in terms of personal development based on more complex self-evaluation and identity processes (Madigan and Munro, 1990; Stern et al., 1999). This broadened conception of identity turns out to be useful when examining green identities.

Green consumer identities are our main focus here. Nevertheless, the green images consumers attach to providers – also in the cases when they are not consumers themselves, are highly relevant. It becomes clear that provider images and consumer identities are closely interrelated. Providers' efforts to make their own green identity to be observed as a green image include attempts to affect the green identity processes of consumers. This can be regarded as one whole set of *activities* of identity and practice on behalf of providers.

The Material Component of Green Identities

The two sectors have material characteristics of significance to the ways green identities are constructed. However, we do not subscribe to material determinism; the material component is only one feature among many. As we will see, the social and political aspects of green identification appear to have higher explanatory value. Nonetheless, when comparing the green identification processes in the sectors the material bases do have a role to play.

As to the *electricity sector* there is no separate green grid, at least not for electricity provided by energy companies. Thus, the principle of green electricity is one of investment in green electricity rather than in a separate green grid. To make the conventional grid greener is hereby the aim of green product differentiation proponents in the electricity sector. This is of course tied to the line-bound nature of electricity; it would be highly inefficient to have more than one main electrical grid system. The *choice* of green electricity is relatively speaking not lifestyle dependent as a continuous activity (Lindén, 1996). Once a household has chosen to purchase green electricity, the continuous routines do not have to differ from the time before the green choice. However in terms of reduced use, electricity is highly lifestyle dependent. Green electricity is instead highly policy dependent, and partly relies on efforts made by providers. Green electricity triggers complex issues of continuous consumer motivation. In contrast to the waste sector, green electricity normally requires an extra fee, which continuously needs to be motivated. Certain companies solve this by distributing information regularly about how the extra money has been invested. This can be interpreted as a way of helping the consumer to build and maintain her identity as a green electricity consumer.

The *separation of waste fractions* deserves a certain bias on the lifestyle and on everyday efforts of households. It is possible to initiate one's own composting without much help from authorities and companies – to be one's own provider. Composting is an ancient practice that gives the person direct ecological feedback; successful composting practices results in odourless soil, useful as a fertilizer for plants. This is certainly true if the person is owner of a house. Even the owner of a multi-occupancy bloc may observe the same effect of feed-back, even though the outcome is more indirect for persons living in the dwellings. In our cases the importance of social feedback from neighbours and providers nevertheless becomes clear. Moreover, queries of identity and policy are raised, especially in multi-occupancy blocs, about the rest of the waste differentiation. What roles do the different incentives by local authority play for household motivation and green identities is important.

Provision and Green Identities

Electricity: How to Make the Invisible Visible

As we all know one cannot distinguish between 'green' and 'grey' electrons in one and the same grid. Nevertheless, energy companies have the possibility of separating the energy sources through organizational and tariff differentiation. An explicit aim of the Energy Company Electra in Sweden is to make the conventional grid greener, and to increase the share of alternative energy offered to consumers.

Still the invisible nature of electricity may lead to socio-material obstacles to having consumers be motivated to choose green electricity alternatives. Consumers are concerned with how their extra money for green electricity is invested. The electricity is mixed anyway in the grid, and 'who knows where my extra money for electricity alternatives goes?'.[9] Thus, the energy providers need to expose the green

[9]At one British energy company this is solved by the company sending out a newsletter every 6 months, so that 'customers are aware of the consequence of their decision' to fund green electricity. British green electricity consumers hold a number of motives: interest in climate change, green pricing, global equity, energy services and concerns over visual impact.

generation, make their green provision and identity visible for consumers. Part of this visualization involves environmentally trusted organizations controlling the green claims of utilities. It is for instance reflected in the ethos of Eco-Power in the UK, namely 'to open up a whole new view of how energy is generated' by improving the amount of Eco-Power in the UK in a very visible way 'so people can see it working'. The informative part of showing consumers how Eco-Power works can be regarded as the manifest function, while a more latent function may be to develop a green identity of the company, transferred to consumers so that they get closer connected to the company on an environmental basis.[10]

This way competition between providers has become more complex and is well acknowledged to involve much more than to simply satisfying the demands of Economic Man. One British Energy Company, for instance, is aspiring to have one of the greenest images of all the electricity utilities in the UK. The company hence emphasizes the green identity of consumers, perhaps even without such identity being very comprehensive yet. Nevertheless, such claims may themselves generate green identity among customers, which they had not developed before they read the slogan from the company. Another tool for green identity construction is that every green customer gets a sticker indicating that the person chooses wind power. Although this appears to be especially important to private companies purchasing electricity a sticker may affect households as well. One effect is simply information to others about the existence of green electricity. The other is a message, not only to others but also to the green consumer herself, of her green identity.

The conspicuous nature of wind power, with its highly visible windmills, makes it efficient in boosting the green identity of energy providers. At the Energy Company Electra in Sweden they believe that wind power has not been a great economic investment directly. Yet, they hold the windmills Annica and Beatrice to have given the company positive publicity merely by being visible. The Swedish company's involvement in a wind power co-operative has strong symbolic value.

An environmentally decisive factor is how much of providers' green image is coupled with actual market shares. The actual percentage of green electricity provided by energy companies is likely to become an increasingly important basis on which consumers will select their providers. Not all energy companies are happy to reveal their green share. At the British energy company they have conducted a survey indicating that 30 per cent of the consumers said they would be happy to buy green energy. Still, at the moment only 25 per cent are actually

[10]The British company was the first one to offer customers a green tariff and was also the first one to get ISO1401. In contrast to many energy companies, especially in Sweden, the British company has decided to target domestic customers, rather than companies, for green tariffs. They hold the reason to be that there is 'too much competition in large companies'.

willing to pay, but '25 per cent of 3 million is a lot of people'.[11] However, there are other processes than the energy companies trying to make their green electricity provision as visible as possible in order to maximize the green share of consumers. In the Swedish case, one might ask why the Electra Company does not build further windmills, solar panels, or initiate other green schemes. Currently they sell 1.5 MWh electricity from wind power, plus the wind power that the company buys from other companies.[12] It turns out that the market price of wind power is so low that it does not cover the production costs. In Sweden, at least, wind produced electricity is thus, on a narrow and short-term basis, economically a bad project.[13] Perhaps this is a reason why the Swedish Company Electra does not market green electricity more actively. Going back to the issue of mixing all electricity sources in one grid, this mixing might make it less visible that green alternatives constitute a rather marginal portion of all electricity generated by the company.

[11]One must keep in mind the gap between these survey results and the green consumer practices. In 1997, the British company said that they would generate 10 per cent renewables by 2010.

[12]The Energy Company Electra in Sweden claims that they have sold all their wind produced electricity. Furthermore, they are engaged in other wind power companies, and have bought shares to be able to offer customers wind generated electricity also when all their own produced electricity is sold. The Swedish Company Electra has started a daughter company. The reason is that they needed a company just to buy and sell electricity from wind-power. Questions emerge as to what extent, and by what means the company tries to turn more households into green electricity clients.

Green Electricity as a commodity has infinite flexibility in the sense that suppliers can order as much green electricity from other producers as is demanded by the clients. This immense *potential* is however not sufficient in order to augment in a sustainable way the ratio of green electricity compared to electricity using 'greyer' sources. Green marketing by energy companies as well as information and greener fiscal policies created by the political authorities are a few of the critical factors.

[13]During the first years of the windmills Annica and Beatrice there were no subsidies. When the third one was built there was however an investment aid of 35 per cent. In addition, they got an environmental bonus, that is, wind power plants are subsidized by the sum that corresponds to the energy tax: 15.2 per cent as of today (fall of 1998). Yet, the investment aid has been reduced today, so that if one builds new windmills today, it will be 15 per cent more expensive. Meanwhile the costs for investing per kWh have been reduced.

The question is if wind power can push out some other production source from the system, perhaps not mainly nuclear power, but hopefully fossil fuels, they hold at the Swedish company. They are not planning to establish new windmills, since, as they hold, the wind conditions are not the best ones in their local area. Nor are they planning to initiate any new co-operative organisations in the near future. The policy they had was that they want to do what they can to share their knowledge and experiences. This lack of initiative of the largest local energy company once they have established a conspicuously green production perhaps ought to be analysed in the context of the large number of limited environmental experiments and projects, which quite rarely are expanded to constitute normality.

On the other hand, the fact that wind produced electricity can be unprofitable for the company is also something that can strengthen the company's green identity: 'Despite the bad economic side of wind power they produce it!'. While green electricity may be unprofitable in the short run, it is fair to believe that it will be advantageous to acquire a solid green company image in a longer perspective.[14]

Faced with certain short-term economic disadvantages, supposedly turned to long-term advantages catalysed by a green identity, energy providers need to make consumers' economic disincentive acceptable. The word choices in the marketing of green electricity become crucial. In October 1997, the UK Energy Company began to discuss how they could satisfy people's rising aspirations and 'give people some of what they want'. They came up with the Green Tariff, although one of the interviewees explains that this is now referred to as the Green Pricing Scheme, because calling it a tariff 'gives the wrong impression'. Later she says that she is sorry that it is called the green tariff now, because 'it's not fixed' ... 'We're trying to give the idea of flexibility so you can adapt how much you pay in.' It is imperative to have the consumers feeling that they belong to a special category of clients. This is done by making green consumers belong to a certain scheme rather than merely pay a different tariff. The phenomenon of strengthening consumer identity by defining them to a special appointed group becomes very obvious here. The flexibility that is stressed in the UK case has interesting implications. Flexibility may work as making price differences less apparent to consumers. Moreover, flexibility, being the basis for charity of various kinds, has strong connotations of benevolence, which might make consumers feel at ease when paying extra, at the same time as their green identity is reinforced.

The moral aspect of green electricity is reflected in the British company's information to their clients where their extra money goes. At the moment, for every £1 put into the green tariff the company also contributes a £1 to spend on 'independent projects'.[15] There was scepticism among the green people lobby about what this would be used for. It resulted in formation of a group of six trustees who talk to individual customers about what they want money spent on. The trustees thus

---

[14]The company's own green identity is kept up on a regular basis, not least through their own maintenance of the windmills. They use Vestas windmills from Denmark. While green electricity may be regarded as somewhat of a fit-and-forget system for consumers, the Energy Company has to maintain the windmills at least once a month. The extra service they buy from the provider of the mill. People come from Vestas twice a year to change oil and make adjustments. Aside from that, the Swedish Energy Company Electra has small costs insurance and administration. They have chosen old and safe technique, rather than technical innovations.
[15]In relation to pricing when they set up the tariff, the British Energy Company originally put the premium at 10 per cent, but then dropped it to 5 per cent. Still, most consumers tend to suggest 10 per cent.

ensure the impartial and democratic character of the benevolent efforts by providers and consumers together. This is a way of acknowledging the social and relative nature of constructing a practice as environmentally sound. To keep green identities of both providers and consumers alive continuous feed-back probably makes the reflexive client draw parallels between this green charity and, for instance, charities to third World countries, victims of earthquakes and war. A certain benevolent competition may emerge as to where the extra money is best spent.

Waste: To Make the Visible 'Doable'

Differentiation in the waste sector takes place as fractionating of recyclables, new choices of collection frequency, and different tariffs. It becomes obvious that successful recycling is dependent not only on benevolent households, but also on active providers, local authorities and companies. Among authorities in the public and private sectors, comprehensive efforts with recycling have become somewhat of an environmental symbol. In Sweden, for instance, the official competition for the title 'the environmental municipality of the year' takes recycling schemes as part of green identity into serious account.[16] As opposed to the electricity sector, the waste sector does not face service providers with competition. Nonetheless, the providers' economic advantages of successful recycling cannot be overestimated.[17]

The waste sector is unique in its revealing how household action can create green identity. In household practices of waste management, providers' recycling schemes have frequently been action-oriented rather than merely appeal to people's green values and attitudes. Several Dutch municipalities, among them Barendrecht, have introduced highly priced refuse sacks as a means of tariff differentiation. To dispose domestic waste, citizens are obliged to use the sack. By increasing the price for sacks and lowering the monthly levy at the same time, the charge for waste collection is related to the amount of waste disposed. Household consumers may save money on waste collection by producing less waste or separating their waste. The option of choosing bin size and number of waste bins in residential areas can be regarded as creating a similar incentive to reduce waste amounts. Among the waste differentiation case, this is one of the top-down ones. Since it is so specifically oriented towards number of sacks and economic incentive, it is hard to see how it

[16]Legal requirements of recycling put on municipalities and materials companies, (DOMUS national policy reports, 1998).
[17]Reduced transportation costs, better quality of the recyclable fractions sold on the market, the avoidance of opening further waste sites, are only a few of the advantages of successful recycling and waste management. On the other hand negative factors can be identified as well. Less waste lead to over capacity in destruction plants which may lead to higher fees for consumers.

would help to create a more solid green consumer identity. Another differentiated tariff project in the Netherlands points at the risk of the pecuniary factor getting too much place in the providers' attempts at motivating households. The experiment covered weighting of both organic waste and the remaining fraction. Wheel-bins were equipped with a chip that identified the owner.

A project with slightly more of household initiatives is the Anniro project in Southern Sweden. The project was connected to a comprehensive program influenced by local Agenda 21 ideas. After making a request to the local street office, households in houses may have their waste collected every other week instead of every week. This presupposes that the households produce low enough amounts of waste, something that the local street office occasionally controlled visually.

> The pedagogical work is about persuading the customer that longer waste collection intervals are not as frightening as many people suppose. Nothing particular happens to the waste if it lies in the bin one more week. But it is crucial to give households the opportunity to separate their waste in a reasonable way. This is partly our responsibility. (Head of the waste section, the Local Street Office).

The waste collection program started when the municipal Street Office contacted the chair of the housing association. Representatives presented a suggestion and asked the residents if they would be interested. The compost model was already chosen. A rotating compost was given to each household for free by the Street Office. The households got a clear idea of what their practices would involve. All three kinds of differentiation were involved: tariffs, fractions and collection frequency. The initiatives in this project hold somewhat of mutuality between authorities and the housing association.

Consumption and Green Identities

Electricity: Free-Floating Greenness

Product and tariff differentiation (PTD) is about consumers creating and being offered new choices. Such processes set social reflexivity in motion. It does not just make citizens welcoming every new choice. It also leads to questioning of how the *system* of product and tariff differentiation (PTD) could become better.

Queries often include how the green responsibility ought to be divided between authorities, providers and consumers. In the Netherlands, the product of green electricity was initially received with scepticism, especially among the environmental and consumer coalitions. There were debates about the link between the commitments that energy companies had made and green electricity. Were consumers actually asked to pay for the commitments that energy companies had to comply with anyhow through regulation? Other consumer reactions were that green electricity schemes represent the opposite of the polluter-pays-principle. The green, conscious consumer has to pay extra for sustainable energy, while the fossil energy users are cheaper off, despite the environmental costs of their pollution. In this way, green electricity was perceived as the servant of a liberalized market in which the government cannot determine the share of sustainable energy generation anymore (Chappells et al., 2000). Whether or not energy companies will use sustainable resources, it is argued, will be dependent on a small group of environmentally conscious and wealthy consumers. In the context of green identity, such system is likely to strengthen the green identity of a smaller fraction of consumers, partly because they make an extra effort in terms of absolute green expenses, and partly because they simply distance themselves from the majority of consumers.[18]

Aside from the political irritability that has emerged among consumers, the reflexive character of money appears to play a decisive role in the green identity processes of tariff differentiation. Georg Simmel (1903/1978) pointed out the indifferent, objective form of money. The purpose it serves can be everything from the noblest pursuits to the most primitive desires. If we dare to label green identity a noble pursuit, product and tariff differentiation helps people become more flexible in their reasoning about what is the noblest. A clear sign of this is the fact that the persons organizing green electricity at the Swedish Energy Company Electra are sceptical to the scheme. In the interviews, they reveal that they prefer, if anything, to buy shares in a wind power co-operative rather than to choose green electricity produced by their own company. Actually, the interviewees do not think that any employee at the company chooses the other way around:

> I can look at myself. One cannot separate electrons in the grid, at the same time as I can understand that an involvement in a certain production type perhaps would lead to that more wind power plants were established, if one is very interested in this. But as to my private financial situation, I'm not ready to provide money for possible wind power plants. And I think many people share this point of view.
> (Person responsible for green electricity, the energy company in southern Sweden).

Moreover, both persons hold that they would rather spend their green money in a wind power company far away than to spend it on green electricity in their own

---

[18]Currently, after the Dutch abolishment of eco-tax for green electricity, consumer prices vary between 2 and 5 cents per kWh, which makes up a total of 50 to 150 guilders per year for the greening of an average electricity consumption of 3,000 kWh/year. It is expected that this will result in a further increase in green electricity consumption in the Netherlands.

company. It reflects both environmental and economic rationality. A wind power co-operative would perhaps produce economic surplus, which could be reinvested in green projects. Such arguments could even transgress the environmental sphere and lead to debates over if one's money could not be used in more urgent projects than green electricity generation.

However, energy companies have found ways to stabilize green identities among consumers. At the Energy Company Electra, for instance, there is the option of writing a contract for 3 years. Contractors can purchase wind power electricity at the price of 23 öre/kWh, which is actually 2 öre cheaper than conventional electricity and 6 öre cheaper than wind power without a contract. The three-year wind power contract is a green identity booster, despite the fact that the electricity actually is cheaper than conventional electricity. The green commitment has identity value. Furthermore, if one wants to change to this company from another one, one during the first two years of deregulation had to get the hourly electricity meter at a price of approximately 2500 Crowns, if one lives outside the 'billing area' of the company. The options, created by liberalization, to purchase electricity from companies in other local areas than one's own, can also function as a way of strengthening identity construction through electricity consumption.

Finally, we have seen interesting examples of how the green identity can turn inward, and raise questions on the daily routine of one's own household. This is a process much in contrast to the political one reflecting over what societal levels ought to have the main environmental responsibility. The green identity directed inwards might ask: 'How can we in our household change to green electricity and still avoid the extra costs?' The answer is simple, but may require certain lifestyle changes: use less electricity.

Customers of the British Energy Company were asked why they choose to get Eco-Power. One of them answered that she signed up to show her trust, but that the electricity bill is in her partner's name. His response was 'why not'. Her partner also noted that the costs were going down, and they get five per cent discount for paying by direct debit. The size of their electricity bill has not changed much, despite the extra 10 per cent for green electricity. The extra fee here works as an incentive to reduce electricity consumption. It should be noted that this mechanism is not always obvious. We have also seen more expansive ideas, implying that: 'the more green electricity we use, the better for the environment.'

Waste: Individual versus Collective Green Identity

In an earlier section on waste we explored the importance of provider initiatives for successful domestic waste management. Here, the households' integration of the green tariff differentiation in individual lifestyles is in focus.

The Individual Identity as Part of the Collective Green Identity

The individual versus the collective can be interpreted in parallel with anonymity and feedback towards a green social identity in the waste sector. Let us start with Anniro in southern Sweden as an initial example.

A few households in Anniro hold that the trial period ended too abruptly (Klintman, 1996). From having provided continuous feedback it has subsequently become more difficult to get help from the Street Office. Some unclear matters refer to how to adopt the composting routines to different seasons. Each season provides the households with certain composting problems. During summers the risk for odour is higher, whereas the compost material is likely to freeze in during the winter. However, when the information was less intense from the municipality, the neighbours began to help each other and give advice. This has lead even households with children in the 'diaper-age' reduce the frequency of garbage collection to every other week.

The Anniro waste project illustrates the two sides common in schemes with waste differentiation. The limited size of the residential area of Anniro makes the neighbours get a rough sense of which households are less successful in their composting and waste reduction. Problems, errors and 'laziness' are in focus when the waste project is discussed formally or informally in the neighbourhood. Resistance to a green identity is more visible than acceptance of a green identity.

In terms of recycling results, the social pressure of a green collective identity is productive. In several projects the same tendency have been noticed. Similar results are found in Bath and Northeast Somerset in the UK (Chappells et al., 2000). The project has developed as a partnership between the local authority and Avon friends of the Earth. It became clear that mini-recycling centres where more successful than larger scale solutions of recycling. Large-scale solutions frequently make the individual consumer feel anonymous, which reduces the green collective identity (Klintman, 1996).

When discussing motivation and morale, the role of economic recycling incentives should be brought up. How do economic incentives of tariff differentiation combine with green identity construction? Economic incentives should not be over-emphasized when trying to stimulate people to adapt green practices. The risk is high of having the economic part over-shadow the ecological benefits. Anniro reveals the positive, value of economic incentives. Money has its importance, but it does not appear that the saved *amount* of money is crucial.

> I don't think that the 500 crowns were very important. It was more that yes, it was good at the same time to be able to save a little money. But at the same time, it would be irritating if two-week collection would be as expensive as one-week. If you have paid a full fee they should come and collect the waste each week. That's probably how most people see it. In that sense money matters, but not in crowns and oren.
> (Man 45 years old, living with one adult and two children in Anniro).

The economic incentive may impact the recycling morale. Yet, thinking about the time and efforts people spend on recycling and learning about composting, this is an illustration, not of Economic Man, but rather of Ecological Man, creating and developing a green identity.

From an Ecological Motive to Broadening of Green Identity

The waste differentiation cases are to a large extent stories of success. This gives much hope to the future of green waste differentiation.

Regarding Anniro in southern Sweden, the head of the local street office maintains that the vast majority of the households are satisfied with the recycling routines and composting. The chair of Anniro is very pleased with how the composting scheme turned out, both in his own household and in Anniro as a whole. He claims that 75 per cent of the households have moved over to a two-week interval of waste collection. The chair of the local Street Office tells us that one household has contacted the street office to buy another compost to use out in their cabin. Yet, there are some households where the composting does not work very well. For them the ecological feedback is sometimes more negative. The one-person households have difficulties getting the compost to work properly. A certain amount of wet waste seems to be needed. The smallest households interviewed in Anniro revealed their sense of being such a marginal part of waste production that composting and recycling from them would be somewhat superfluous. However, the increasing share of single households in Europe (in Sweden approximately 50 per cent) makes the impact of their recycling obvious. A plausible solution would be to have a few households share compost equipment.

In Bath in the UK the comprehensive recycling scheme with direct feedback to residents are considered successful according to providers. The wide range of materials collected from residents is separated into special multi-compartmentalized vehicles. Such sorting permits direct feedback to residents. Materials that are not accepted for recycling are left behind in the box with a note explaining why they cannot be taken. Today the green recycling scheme covers about 45,000 households in the Bath and Wansdyke area and employs 23 staff members. Residents in these areas recycle or compost over 25 per cent of their dustbin waste. Accommodating the needs of different residents encouraged a higher participation rate and sorting allows for maximum recovery and minimum rejection of recyclable materials, as well as reducing the need for post-collection sorting.

The Dutch example in Oostzaan is also very promising. They included a reduction of remaining waste of 60 per cent and a reduction of green waste of 50 per cent compared to 1992. It was clear that consumers are not used to the idea of paying for their waste. The period of October 1993 until April 1994 proved that application of financial incentives discouraged the production of waste. Research showed that two-thirds of the inhabitants were content with the system as on average they had to pay less for the removal of domestic waste. A small part of the remaining waste fraction is brought to other municipalities, but it is supposed that this side effect will decrease as habituation occurs.

One of the virtues of green identity construction, as opposed to mechanically adopting a new behaviour pattern, is that it can lead to broadened reflections of how to green one's lifestyle. Changes of one environmentally related action might lead to other action changes, not only for environmental, but also for practical reasons. In Anniro there are clear practical motives for reducing the amount of packaging of the goods that they purchase. The composting has led the households to plan and adopt their consumption after the limited space in the garbage cans. The single household's results are visible:

> Now when we have changed to two-week collection it means that we buy different things than we used to. We avoid large detergent packages and marmalade jars. Now I buy refills that there is room for in the garbage can.
> (A resident in the housing area, Anniro, southern Sweden).

Similarly, Horsham in the UK is now delivering the 140-litre bins free of charge to every household in the district, while households requesting the larger 240-litre bin have to pay an administrative and delivery fee. One of the main outcomes was that the small bins and recycling baskets provided households with the incentive to recycle and compost rather than throw waste into the bin.[19] The composting practice constitutes an extension of the households' greening practices.

Conclusions and Discussion

Although we distinguish between consumers and providers, it should be noted that product and tariff differentiation (PTD) can be provided by the consumers as well. The theoretical frame of reference is relevant to identify two important processes in greening provision and consumption of goods and services in public sectors, namely the process of identity formation by consumers and providers and the process of materializing public goods and services. Both processes include strong relations between consumers and providers.

In the interpretation of qualitative interviews four aspects of identity formation were identified closely related to nature, environmental rescources, economic factors,

[19]Extract from the Audit Commission Report, *Waste Matters – Good Practice in Waste Management*, 1997.

Identity				Materialization of products		
Ecological Identity	Environmental identity	Green economic identity	Green social identity	Make visible	Make acceptable	Make doable
Electricity +	+++	+	+	+++	+++	*
Waste +++	++	+	+++	+	++	+++

+ indicates the relevance of identities and materialization aspects in sectors.
*not relevant for electricity.

Figure 1   Forms of green identity and materialization of products and services in waste and electricity sectors by product and tariff differentiation (PTD)

as well as factors tied to social relations. Green identities among consumers and providers can be divided into:

• ecological identity;
• environmental identity;
• green economic identity;
• green social identity.

In figure 1 we show how consumers in each sector may hold more than one kind of identity. Also, the power of identities may differ (marked with a number of +). Identity through consumption is much more than just a question of use-values in the narrow, utilitarian sense or merely a question of prestige. All four types of identity presented are closely tied to personal processes of self-evaluation.[20]

Identities

Ecological Identity

Ecological identity refers to awareness of local and small-scale eco-cycling, as well as of saving resources. It often involves subtle perceptions of feed-back from ecological processes. Successful composting practices, for instance, are indicated by the quality of the soil. Reduced waste going to landfills as a result of product and tariff differentiation (PTD) can sometimes be seen in local figures of waste reduction.

Environmental Identity

Environmental identity pertains to the more intellectual awareness of large-scale environmental problems debated in media and researched in the advanced sciences. A few of these problems are global warming and ozone layer depletion. Environmental

[20]For an in-depth discussion of consumer identity and image, see Warde, 1994.

identity turns out to be most intimately connected to consumption of green electricity, contributing to a somewhat larger share of investments by greening the grid. Waste differentiation is also related to environmental identity, as it has been a substantial part of the think-global-act-local campaigns. Reduced waste incineration has also been related to larger issues in the public debate.

Green Economic Identity

Green economic identity is best illuminated by the phrase: 'We save money on it, and it is good for the environment, too.' Green electricity consumption is the least connected to such a statement, since it is an economic disadvantage for households to purchase green electricity. On the other hand, this economic disincentive can sometimes function as strengthening the environmental identity. In waste differentiation schemes, the green economic identity has been reflected in several interviews. However, a symbolic economic advantage appears through the cases to be more crucial than the actual size of the economic gain.

Green Social Identity

Green social identity, finally, emerges in cases where informal and formal, political efforts have an essential role to play in developing and implementing product and tariff differentiation (PTD). The waste sector presents several examples of this. Continuous feedback between providers and consumers has been a key to success. Green electricity is largely based on communication between provider and consumer. It remains to be seen whether or not social networks of neighbours and residents can be more developed.

Materialization of Products in Public Sectors

Provision also contains green identity-creating components, which have to be noticed by consumers. We have divided the materialization of products into:
- make visible;
- make acceptable;
- make 'doable'.

Make Visible

Providers in our electricity cases put their main efforts into making their green electricity (by nature invisible) conspicuous. This serves the dual purpose of marketing the product, and establishing the green identity of the company. Interestingly, conspicuous slogans of what today's consumers 'really want' may help construct green environmental identity among consumers, whom may not have thought much about environmental issues before. Moreover, making visible the fact that green electricity production is not always economically profitable in the short run also reinforces the idealist image of the company. The main aim of providers of green electricity is to make it visible on the market, hence supplying environmental identity on sale and strengthening the green environmental identity of consumers. Waste, on the other hand, is visible. The problem is rather to make the complex composition of fractions visible for consumers.

Make Acceptable

Three crosses in making electricity acceptable refers to the economic disincentive of green electricity, which actually made even providers themselves refuse to accept this disincentive in their role as consumers. Two crosses in making waste acceptable reflect the odour and hygiene aspects of waste differentiation, to keep several bins in the house, to compost and rinse organic waste. However, these concerns are usually turned into acceptance if only the feedback from providers and neighbours is sufficient.

Make 'Doable'

This has to do with the physical and practical conditions that providers are largely responsible for, whether it be the consumers themselves or strictly providers. We want to bring forth an action-oriented approach. Providers have two main tasks here, which relate especially to waste differentiation. One of the tasks is to develop policy and management systems in the sectors so that consumer practices become doable. In the waste sector the number of waste fractions that are recycled by providers is fundamental to what consumers will do. The second task is to improve the physical and practical conditions for waste differentiation at the lifestyle-level among consumers. The distance to bins for households, the level of physical preparedness in the households, information about how to decrease practical obstacles at an everyday level are essential factors.

We have seen examples of how consumers and providers co-construct different forms of green identities. In figure 2, examples are given of product and tariff differentiation efforts among consumers and providers in waste and electricity sectors. It is shown how consumers take active part in provision. Nevertheless, the differences between sectors are obvious. As to green electricity, this is the sector where consumers are most distanced from provision, other than in terms of economic provision connected to green electricity projects. The waste sector has much more of consumer activity in providing the conditions for practice together with providers. The waste sector is the one where consumers most independently have the possibility of improving conditions for product and tariff differentiation (PTD) practice.

	Consumers	**Providers**
Electricity	• Selling green electricity provided by consumer coops to energy companies	• Providing technology • Tariff differentiation (disincentive) • Make visible
Waste	• Suggestions of number of fractions • Physical/practical provision of compost equipment and household bins for recyclables	• Policy conditions, • Number of fractions • Improve practical household conditions • Tariff differentiation (incentive) • Make doable

Figure 2   Co-provision of product and tariff differentiation (PTD) of consumers and providers

Several policy-relevant matters have emerged. One is the importance of stimulating providers in the utility sectors to establish better collaboration with consumer groups and grass-root organizations. It has become obvious how often providers base their level of differentiation on over-simplistic assumptions about consumers' preparedness or willingness to act. However, in improving the collaboration with consumers it is not sufficient to increase the number of surveys about the extent to which consumers are willing to act in accordance with certain levels of product and tariff differentiation. Through studying cases of waste the strength of the practical process towards acceptance and making things doable became clear. It is after having tried a practice that consumers acquire ecological or environmental identities to the extent that they take further initiatives, make suggestions to providers, and broaden their green practices. Still, this is only one side of the coin. More fundamental appears to be the dual tendency of utilities becoming increasingly bigger, whereas product and tariff differentiation (PTD) might better benefit from several emerging, small, flexible companies providing specialized green choices. Better green market segment studies seem to be needed.

Note

Method: Qualitative interviews with persons in the waste sector (35 households, 3 persons from the management in three housing organisations, 3 persons from the municipality organization representing the waste sector) and the electricity sector (15 households, 3 persons in management positions in the Electra Energy Company, 2 persons from the Wind Power Co-operative) in Sweden. The sampling of households was based on theoretical arguments, that is to say the chosen households were assumed to represent different categories of households, i.e. young/old households, with/without children, and with variations in number of members. The interviewed representatives from providers were all in positions responsible for policies, planning, organizational matters or marketing. The Swedish interviews as well as parallel interviews in The Netherlands and United Kingdom were parts of the empirical material in the EU-project *Citizenship, Public Utilities and the Environment*. The EU-project is reported in Chappells, H., Klintman, M., Lindén, A.-L., Shove, E., Spaargaren, G., and van Vliet, B., (2000), *Domestic Consumption, Utility Services and the Environment*, Wageningen: Universities of Lancaster, Wageningen and Lund.

References

Campbell, C., (1995), 'The sociology of consumption', in D. Miller (ed.), *Acknowledging consumption*, London: Routledge, pp. 96–126.

Chappells, H., Klintman, M., Lindén, A.-L., Shove, E., Spaargaren, G., and van Vliet, B., (2000), *Domestic Consumption, Utility Services and the Environment*, Wageningen: Universities of Lancaster, Wageningen and Lund.

De Young, R., (1993), 'Changing behavior and making it stick. The conceptualization and management of conservation behavior', *Environment and Behavior, 25* (4), pp. 485–505.

Dobson and Nicolaidis (1974), 'Preferences for transit service by homogeneous groups of individuals', *Proceedings of the Transportation research Forum, 15*, pp. 201–209.

Dobson and Tischer (1976), 'Beliefs about buses, carpools, and single occupant autos: A market segmentation approach', *Proceedings of the Transportation Research Forum, 17*, pp. 201–209.

DOMUS national policy reports (1998), Wageningen Agricultural University, Department of Sociology, Wageningen.

Dwyer, W.O., and Leeming, F.C., (1993), 'Critical review of behavioral interventions to preserve the environment research since 1980', *Environment and Behavior, 25* (3), pp. 275–321.

Geller, E.S., Winett, R.A., and Everett, P.B., (1982), *Preserving the environment: New strategies for behavioral change*, New York: Pergamon.

Joas, H., (1996), *The creativity of action*, Chicago: Chicago University Press.

Johansson, B., (1997), *Stadens tekniska system: Naturresurser i kretslopp (Technical systems in cities: Circulation of natural resources)*, Stockholm: Byggforskningsrådet.

Klintman, M., (1996), *Från trivialt till globalt (From the 'Trivial' to the Global – Deriving Environmental Influence From Motives and Practices in Urban Realms)*, The research group Boende och bebyggelse, Department of Sociology, Lund. (In Swedish, abstract in English.)

Lindén, A.-L., (1996), 'Från ord till handling (Words versus Action)', in L. J. Lundgren (ed.), *Livsstil och miljö: Fråga, forska, förändra*, Stockholm: Naturvårdsverket förlag, pp. 67–99.

Madigan, R., and Munro, M., (1990), 'Ideal homes: Gender and domestic architecture', in H. Putnam, *Reason, truth and history*. Cambridge: Cambridge University Press.

Rosén, B., (1988), *LortSverige: 50 år efter Lubbe Nordströ. (Dirty Sweden: 50 years since Lubbe Nordström)*, Stockholm: Svenska Kommunförbundet.

Sayer, A., (1979), 'Epistemology and conceptions of people and nature in geography', *Geoforum*, vol. 10, pp. 19–43.

Simmel, G., (1903/1978), *The Philosophy of Money*, London: Routledge and Kegan Paul Ltd.

Skinner, B.F., (1971/1990), *Beyond freedom and Dignity*, New York: Bantam/Vintage.

Stern, P.C., et al. (1999), 'A Value-Belief-Norm Theory of Support for Social Movements: The Case of Environmentalism', *Human Ecology Review* vol. 6.2, pp. 81–97.

Swedish Waste Collection and Disposal Act, 1979:596, Section 2a, 1990:235.

Wachs, M., (1991) 'Policy Implications of Recent Behavioral Research in Transportation Demand Management', *Journal of Planning Literature*, vol. 5.4.

Warde, A., (1992), 'Notes on the relationship between production and consumption', in R. Burrows and C. Matsh (eds), *Consumption and class: Divisions and change*, London: Macmillan, pp. 15–31.

Warde, A., (1994), 'Consumption, identity-formation and uncertainty', *Sociology, 28.4*, pp. 877–898.

Markets, Business and Sustainable Repositioning

Rolf Wolff and Olof Zaring

Introduction

Two decades ago, environmental issues were of little concern to most business organizations. In the 1990s, the environment started to become an important factor in managers' decisions. The number of environmental regulations increased and environmental non-governmental organizations that did not exist in the 1970s now have large numbers of members. Many European countries have green parties with parliament representation, and the general public claim to be increasingly interested in environmental issues (Bennulf, 1994). Business magazines have special issues on the environment, the 'environmental manager' is an established profession and corporate environmental management is on the curriculum of an increasing number of universities. Managers are exposed to environmental issues more frequently than earlier, and experience pressure to become more environmentally considerate.

In the past the bulk of environmental research has been in the realm of engineering and natural science. Although sometimes showing conflicting results, this research has created awareness that many environmental problems are the result of human activity and that it is necessary to find appropriate solutions. Such research has come up with descriptions of the state of the world and presented potential solutions. However, these solutions are often technical in nature (Carlsson Reich et. al., 2001), and are not adjusted to the social context in where they have to be implemented. Organizations are not strictly rational, and the 'good' solutions that technical research produces are simply not put to action. Implementation in the business context is one of this century's great challenges.

When environmental issues become management concerns, it is important to keep in mind that managers may simultaneously need to handle the pressures to be efficient and adaptive (March, 1994). Efficiency pressures involve short-term financial survival and improvements of existing routines. Adaptation, on the other hand, concerns learning (i.e. the organization's ability to replace strategies and knowledge that were earlier taken-for-granted with new ditto). In the environmental context, managers need to treat pressures much the same way. Efficiency expectations can be treated within existing modes of conducting business. Cleaner technologies, environmental management systems and environmental information are examples of instruments that may deal with efficiency criteria. Adaptation issues may in addition involve strategic issues such as new product development, or functional sales rather than product sales. It is thus conceivable that the components of the 'environmental challenge' to business organizations change over time.

Environmental problems are complex in the sense that any individual actor in society cannot solve them. They require multi-disciplinary knowledge, and involve strong emotional and value-laden elements. The environmental problems are one cause of modern society's uncertainties and risks. Individuals' daily collective actions may influence the environmental situation in other parts of the world (e.g., a choice not to buy organically grown agricultural products).

Issues of values and emotions are thus becoming increasingly important to business organizations. Such 'soft' factors are becoming legitimate business concerns, as societal pressures demand so, and 'soft' issues are incrementally translated into 'hard' business facts. Knowledge of this change is something that natural scientists cannot provide. They give us the crucial knowledge regarding the state of nature, but not an understanding of how individuals, in their respective roles, as part of varying organizational contexts, see their own impact on the natural environment.

The last five years or so, the realm of the environment has been broadened by the term 'sustainability'. As over 50 of the world's 100 largest economies are not states but companies, and as 85 per cent of the world's development resources are controlled by the business world, the issue of 'corporate social responsibility' is added to the environmental responsibilities of business. Some view environmental concerns as an inherent part of companies' social responsibility. Whatever the conceptual definition, the boundaries of what companies' responsibilities are, are in flux and increase the uncertainty amongst business leaders as how to deal with these challenges, in any business context.

This contribution sets out to discuss this development, with a focus on the business level. It builds on extensive research during more than a decade attempting to cover what the 'environmental' and 'sustainable' business challenge is about.

In part one we discuss the relationship between consumer and industrial markets and sustainability as well as the relationship between the latter and the role of the financial market.

In part two we present our theoretical base and conceptual model that guides our research, when it comes to the repositioning of companies, due to their attempt to adapt to environmental (and social) demands. We illustrate our concept of 'sustainable repositioning' within business ecosystems with two cases from the oil industry: BP and Shell. It should be noted that the eco-system concept used here to denote a theoretical approach originates in organizational analysis rather than the biology, (cf. Hannan and Freeman, 1989). A *business ecosystem* consists of similar and dissimilar *organizational* populations kept together by mutual interdependences. Each population of units (companies) has its own dynamic processes of competition and cooperation, and the members of a population draw on similar resources, including natural resources. A re-positioning by a business population may occur as the result of changes in the composition of the resource base for that population.

In the final part an outlook regarding the research agenda during the next decade is made, and includes our thoughts regarding the emerging social agenda.

Markets and the Environment

Many regulative and management instruments have been developed during the last decades in the attempt to accomplish improvements in the environment. Since the Stockholm Environmental Conference in 1972 a strong focus has been on regulating for the caring capacity of the globe. Daly (1992), an environmental economist, for example has formulated some of the principles that are necessary in order to guarantee what he calls the 'Economic Plimsoll Line'. He defines this 'plimsoll line' as an instrument to avoid environmental 'overloading and sinking' (Daly, 1992, p. 189). Keeping this and other macro-economic environmental frameworks in mind, we are concerned with approaches that meet with such normative requirement in theory and practice, and which describe the context of 'management processes'.

The important elements in the attempts to achieve sustainability have been:
- Regulation;
- Consumer awareness;
- Companies' solution of end-off-pipe problems;
- Companies' green product development.

The interplay between these elements can be seen in figure 1 on the next page.

In short, regulation has been geared towards the control of physical resources in production and lately also consumption processes (product use). The guiding normative concepts have been that environmental regulation should control business production to reduce pollution and resource use; that consumers should be educated so that they consume environmental products and discredit those products that are less environmentally adapted. Regulation and consumer awareness should

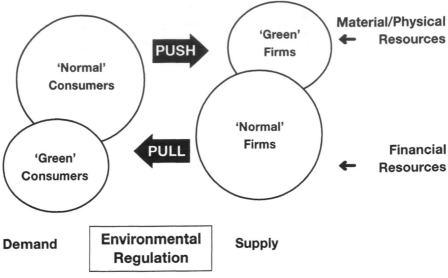

Figure 1 Markets and regulation for environmental control

influence what companies do; companies then would influence the 'pull' in the market through providing adequate products.

The ongoing discourse on 'environmental awareness' in society seems to have indicated that consumers de facto behave according to their expressed values, which would imply that they consume better environmentally adapted products when these are offered in the market. This belief has created some over-optimistic marketing strategies amongst those companies trying to gain 'first-mover-advantages' in a market. Although consumers seem to express positive environmental attitudes, when it comes to consumer behavior the challenges for companies are much more ambiguous. Also, when looking at segmentation and marketing strategies, there is a strong indication that 'green' consumers differ considerably depending on what part of the world we look at. From a business marketing point of view a differentiated marketing strategy is required (Bragd, 1998).

Environmental reporting has been an important issue during the last decade. Its main importance probably has been to reduce uncertainty in the relationship between companies, that is business-to-business relations (suppliers) and thus, in industrial markets. The quest for transparency in relation to the 'push'-side of the market has had two important target groups in mind. On the one side the idea has been to make business more transparent for companies' stakeholders such as customers, non-governmental organizations, governments, employees and suppliers. The other stream of reporting ideas has been oriented towards supplier-buyer relations.

Reporting has been very much adapted to the 'physical-resource-stream' concept, that is, focusing on the environmental impact and performance improvements of a company. Management aspects have later come into consideration through environmental management systems (like EMAS or ISO), aiming at establishing administrative routines for environmental management. Still, a considerable gap exists between the operative environmental management and companies' strategic decision making about business goals and core products and services.

From a more general point of view though, business companies are as much dependent on physical resources as they are on financial resources. Looking at these financial flows as strategic resources, the question of liabilities has to be combined with the issue of innovation and competition (Carlsson Reich et. al., 2001). According to our view the financial market has an influence on the rate of the flow of physical resources in companies. Therefore these markets are of strategic interest with regard to sustainability, from a policy perspective, a business perspective as well as an environmental perspective.

Capital Markets and Sustainability

Capital markets have a potential power to influence what companies do by giving or preventing companies' access to financial resources. One of the strategic questions is if and how financial markets influence resource use and resource efficiency.

Currently two global discourses are dominant in two different arenas. On the one side, we have the shareholders' value concept (Copeland et al., 1994; Black et al., 1998) that has been guiding financial markets the last two decades. Related to the concept is the critique of *shortermism* and one-sided shareholder orientations. On the other hand we have the sustainability concept as formulated by The World Commission on Environment and Development (1987).

Three basic positions can be derived from these two discourses:
- Shareholders' value (SHV) is the foremost objective of a companies' raison-d'être, or as it is sometimes referred to as 'the business-of-business-is-business'. This position implies that all company actions should be sub optimized according to a simple philosophy, expressed in the following statement:
'We raise capital to make concentrate and sell it at an operating profit.
Then we pay the cost of that capital.
Shareholders pocket the difference.'
(Coca-Cola's former CEO Roberta Giozueta; in Black et al., 1998, p. 22).

- The counter position to that *pure* shareholders' value perspective is the concept of sustainability as it evolves in various practices. Here sustained shareholder value is treated from an ethical basis claiming a holistic economic approach including environmental, social and financial objectives.
- The third position is an 'environmentalist' position where shareholders' value is a subordinate category to nature's demands on ecologically adapted economic systems.

The discourses of shareholder value and sustainability contain a large potential for conflict. The critique of SHV claims that it is short term oriented, leads to avoidance of innovation and renewal, which in itself leads to deteriorating profits and also to unrealistic profitability objectives. Strategist Gary Hamel talks about 'Hitting the wall of diminishing returns' and questions the inherent destructive power of the shareholder value concept (Hamel, 2000, p. 37). Shareholder value orientation also implies a neglect of other stakeholders than owners. The sustainability discourse on the other hand can be accused of neglecting business realities, such as shareholders interests and focusing too much on environmental and social demands. The 'environmentalist' position may be seen as the other extreme end of the continuum.

It may seem that both discourses are contradictory, though we believe they are not. We propose that shareholders' value will be best accomplished if companies include environmental and social responsibility objectives in their value creating statements, processes and products. We argue for a holistic business approach, balancing stakeholders' demands and sustainability requirements, which will create sustained success for the company. The balancing process though is not easy and requires an ability of the management of a business to deal with value-dialogues that are conflicting and inherently contain tradeoffs.

One remark and reflection, though, is important here. It seems to us, that the cultural origins of the various discourses have not been objective for analysis. Obviously, the shareholders' value concept is a US-based notion:

In continental Europe and Japan intricate weightings are given to the interests of customers, suppliers, workers, the government, dept providers, equity holders and even society at large. In those quarters, maximizing shareholder value is often seen as shortsighted, inefficient, simplistic, and perhaps antisocial ...

A U.S.-style system based on maximizing shareholder value, accompanied by broad ownership of debt and equity and an open market for corporate control, appears to be closely linked with: a higher standard of living, greater overall productivity and competitiveness, and a better functioning equity market.

(Copeland, Koller and Murrin, 1994, p. 3).

In other words – the U.S.-model, as it has been successful, should be or will evidently be exported through the financial market to Europe and Japan. As a matter of fact, this is what has happened during the last five to ten years. The shareholder-value-concept has swept through European boardrooms and changed not only the rhetoric of management in the European context, but also the structure of European industry.

The sustainability concept on the other hand has Scandinavian – European origins, which might reflect different cultural ideas about business purposes and ethics and about welfare creation and distribution on the societal level. In a cultural anthropological sense, companies are not functional for particular interest groups. Businesses are created to satisfy needs in society and to create wealth. Wealth creation also is a function of social organization and the distribution of responsibilities in society (Thurow, 1999). The foundation of social organization and responsibility distribution is a value decision, based on societies' cultural inheritance. The question of shareholder value thus touches at very deep routed aspects of the institutions that are legitimized to create wealth and its distribution in society.

As mentioned before, we consider financial flows to be of utter importance when it comes to the establishment of sustainable businesses. In Sweden and Scandinavia, the last years much discussion has been devoted to the potential effect those environmental funds may or may not have impact on industries' environmental performance. However, in a recent study of environmental funds we found that these financial products have relatively little impact on companies' environmental performance. Also, consumers very often do not know what fund products they actually invest in. The definition of 'success' of an environmental fund very much depends on those actors in the system that conduct an evaluation.

An analyst may either focus on environmental or financial performance (or both), a portfolio manager might be interested in the size of the fund under management and the frequency of transactions, government agencies might focus on the environmental performance and the consumer might like to have a product 'that does good' for the environment and also give her satisfying return on investment over time. Mere environmental technology funds do not seem (with few exceptions) live up to these expectations and few funds are able to meet all the other objectives simultaneously (Symreng, Wolff and Zaring, 1999).

Aside these more distinct financial products we do see a strong trend in the more strategic financial flows towards the use of sustainability criteria. Institutional investors voice a growing concern regarding the ways in which they manage their assets under management. Pension Funds and other institutions more and more attempt to include environmental and social/ethical criteria in their assignment when choosing an outside manager for their assets. Reliable methods to evaluate companies from a sustainable set of criteria, though, have to be developed.

Two events contribute to the growing transparency with regard to company evaluation and the adaptation of sustainability criteria in asset management. The first one has to do with the increasing awareness of investors that 'sustainable' investments are a matter related to investors' brand value. The other mechanism is the emergence of indexes that put a demand on companies to make transparent how the company manages sustainability.

A study conducted by Butz and Plattner (Sarasin, 1999) claims a 'significant positive correlation between environmental performance and financial returns in environmentally relevant sectors'. The study was not able to establish a similar correlation between returns and social responsibility. We have seen a growing integration of the sustainable pillars (economy, social responsibility and environment) in the creation of new fund products, and/or the strengthening of existing methodologies. The economic performance of companies included in an investment portfolio will thus be complemented with both evaluation of the environmental and the social performance.

Financial markets and its actors have a considerable relevance for the transparency of sustainable business processes. Three conditions are necessary in order to improve sustainable business:

- Sustainability criteria have to be incorporated in the evaluation of companies.
- It is necessary to being able to demonstrate that sustainable management contributes to better economic outcomes (at least long term).
- Benchmarks are needed that cover all emerging sustainability dimensions and these have to be available for the public.

1999, the launching of the Dow Jones Sustainable Group Index (DJSGI), marks a milestone in the process towards better transparency regarding sustainability in the financial market. As the DJSGI and other similar evaluation methods gradually improve, global benchmarks will be available, including single company benchmarks in selected industries. Also, as transparency is improved over time better methodologies with regard to screening processes will be available. The point at this time is not that the Dow Jones methodology is perfect or correct. The point is, that one of the global players in the financial market gives legitimacy to issues that were previously treated as 'soft' or irrelevant. The new index and other similar activities contribute to forcing companies to make transparent, report and evaluate continuously, as well as communicate their measures in the sustainable framework. In more theoretical terms, we are now witnessing a process during which increasing transparency and improved financial evaluation methods will successively improve the evaluation of companies' values. Hidden risks and unknown values will be uncovered and may also contribute to a re-evaluation of shareholder value as a management driver.

Capital Market Research

One strain of research about financial markets is informed by the notion that a financial market has two basic categories of actors, cf. the seminal work of Berle and Means, and the more contemporary developments on agency theory (Berle and Means, 1932; Jensen and Meckling, 1976). One category of actors, the principals (in this case the investors), holds assets that can be freely invested with the aim of increasing the total assets. Such investing is assumed to be made by buying securities, usually stocks or bonds, issued by the managers of companies. The securities are bought on the assumption that they will provide a future return to the holder, the investor, of the security. The agents of the investors, such as company managers, act under the obligation to deliver the returns on invested assets to the principal. They are expected to promote the interests of their investors in all their actions. In real life the principal would be any investor and the agents would be the managers of the firms issuing securities such as shares of equity. Research has often had the aim of developing schemes for ascertaining that managers act in the interest of the shareholders in the company, i.e. that they should have disincentives for diverting assets for personal use or incentives to deliver the returns to the investors. One such scheme mentioned elsewhere in this chapter is 'shareholder value'.

The investor buys and sells shares based on the information provided by the managers of firms and the investor can compare different firms according to his or her preferences. This results in investment decisions to buy the shares of particular firms. Thus the aggregate of the decisions to buy or sell shares set the price on securities by analyzing information about future returns to be delivered by firms' managers. This comprises the capital market.

Now, one might ask what this has to do with the environment or ecological sustainability? The relevance is based on the fact that companies use natural resources to increase the wealth of investors. The managers, as agents of a firm's investor, strive to produce more goods and services at a higher profit margin. This leads to managers' searching for larger amounts of cheap raw materials, less costly production processes, and efforts to increase output prices. Usually, or at least often, this behavior has resulted in a degradation of the natural environment.

In economic theory this 'traditionalistic view' on the role of corporations and their managers has been associated with Milton Friedman (Friedman, 1962, 1970; Friedman and Friedman, 1980). His position can be summarized as follows:

> Business leaders have a prime responsibility to owners of shares to maximize their value. Managers act as agents of shareholders. They have as such no mandate to embark on socially responsible projects, if and when these activities do not contribute to

enhanced abilities to generate firm profits. In addition, managers should not refrain from profitable investments that – of course – should satisfy all legal constraints. Managers own personal social agenda should thus not be confused with their shareholder responsibilities.

In the Friedman world 'the social responsibility of business is to increase profits' (1962, p. 133). In Friedman's terms corporate social responsibility is a 'subversive doctrine': 'Few trends would so thoroughly undermine the very foundations of our free society as the acceptance by corporate officials of a social responsibility other than to make as much money for their stockholders as they possibly can. This is a fundamentally subversive doctrine.' (op cit., p. 133)

Further Friedman's argument says that managers do not have comparative advantages when it comes to implementing social programs. These arguments are, of course, convincing to some extent. We do claim though, that the social agenda since the 60s has changed considerably. Today, more than 50 of the 100 largest economies in the world are companies. More than 85 per cent of the world's financial development resources are controlled by business. Governments are weakened and the corporate world plays an ever more important role for solving poverty, environmental and other social problems.

However, aside the 'traditionalistic view', one class of investors nevertheless have a preference for selecting firms that have a beneficial influence on the environment, and that use resources in a more ecologically sustainable way. Assuming that the capital market evaluates ethical information about companies, such an investor would then analyze the information about resource use and environmental impact provided by firms' managers to select stocks to invest in.

However, environmental information may not be forthcoming from corporations in a way that can be readily used by investors in their decision-making processes. Even if companies delivered such information it is notoriously difficult to interpret. This means that it can be very difficult to establish whether an investment is truly ecologically beneficial to investors.

Thus, the capital market's role and functioning in relation to sustainability is potentially an important one. However the need to analyze complex environmental information makes the market more complex and introduces hindrances and filters in the reciprocal information flows between the principal and agents operating in the environmental market-segment. It might prove difficult to establish whether the agents for 'environmental investors' really act according to the wishes of their principals.

We have briefly touched upon consumer markets, industrial markets and financial markets as a contextual aspect of how managers may or may not react to demands for adapting to the environmental considerations. In the next part we will develop a framework for studying environmental change processes on a company level.

Sustainable Repositioning of Industries

Companies seek positions in markets that enable them to survive. Survival builds on the ability to receive and maintain critical flows of resources and capital. The main body of strategy research and theory has been oriented towards finding models to guide managers in the struggle to maintain and improve these positions (Hedberg and Wolff, 2001).

Environmental challenges have introduced a new dimension into both the strategic practices and strategy research. Our own theoretical reflection goes beyond the instrumental strategy repertoire and selects an organizational theory, that is both evolutionary and has a strong focus on the context in which changes occur.

We attempt to develop a contextual approach. Based on our previous research, we believe that single companies' guiding assumptions about the world only can be understood when the context in which these are developed and disappear is taken into consideration. Thus, our perspective is on the single company as a part of an industrial network and its ecosystem.

From the 'Theory of Business' to Business Models

What business organizations do or not do is an outcome of the assumptions these companies develop over time. Peter Drucker (1994:95) has labeled these assumptions as 'the theory of business':

> The assumptions that shape any organization's behavior dictate its decisions about what to do and what not to do, and define what the organization considers meaningful results. These assumptions are about markets. They are about identifying customers and competitors, their values and behavior. They are about technology and its dynamics, about a company's strengths and weaknesses. These assumptions are about what a company gets paid for. They are what I call a company's theory of business.

To understand companies' actions it is not sufficient to describe and analyze the theory-of-business per se. The theory of business is a narrative that summarizes a business concept; it is a story to be told to external and internal stakeholders. The theory of business a company tells has to be translated into a rational model that describes how desired outcomes and results can be accomplished.

As we have seen during the late nineties, the relationship between the story and the actual business model can be rather weak. Many dotcom-companies have been very successful in telling 'theories-of-business'-narratives and thus generated extensive financial resources. The translation of these stories into actual business has been weak and the extensive death of many of these companies is an illustration of this weak relationship.

A theory-of-business has to be translated into a business model, a rationale that analytically describes the necessary elements, resources and activities that have to be carried out to produce a desired outcome. Usually this is referred to as the 'strategy' of the firm and much normative research has been devoted to support strategizing in companies (Hedberg and Wolff, 2001, for an overview).

From Business Models to Business Systems

The theory is translated into the business model and its manifestation is the business system that a company implements as a whole. The business system can be described in two ways: through its components and its performance. The elements

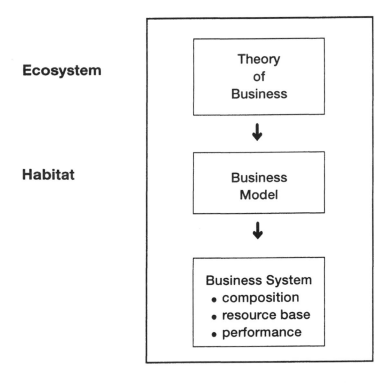

Figure 2   The habitat is the physical context of the business system, whereas the ecosystem is its institutional context

of the business system are activities and relations between those (competitive and/ or cooperative). Performance measures can include financial criteria, but also other criteria such as social and environmental performance.

Organizational Ecology

During the last decade 'organizational ecology' has gained ground in organization theory (Hannan and Carroll, 1992; Hannan and Freeman, 1989). Among all sub-fields in organization studies, organizational ecology has exhibited the greatest theoretical and methodological consensus (Pfeffer, 1993). Ecologists assume that organizational populations form and have unit character in the sense that they respond in similar ways to environmental pressures (Hawley, 1950). Populations are dependent on distinct combinations of resources, called niches, which support them. It is assumed that (Aldrich, 1999:43):

- populations compete for resources within the same environment and that they are therefore in a constant state of competitive interdependence;
- competition pushes organizations toward adopting similar forms, resulting in homogeneity or specialization of forms within different niches;
- organizations find niches in order to defend themselves against competition;
- sometimes organizations make common cause with one another as they compete with other organizations and populations, thus creating a mutualistic state of cooperative relations;
- competitive and cooperative interdependencies jointly affect organizational survival and prosperity;
- and result in the distribution of organizational forms that adapt to a particular environment.

Three levels of complexity are applied in the analysis of ecological processes (Ziegler, 1996, p. 3956). The first level concerns *the demography of organizations*. It tries to relate variations in life cycle processes across individual organizations to changes in the environment. The second level deals with *the organizational ecology of organizations* and analyzes growth and decline as the dynamic outcome of interacting populations occupying partially overlapping niches. The third level refers to *the community ecology* of organizations. A community is an organizational system with interacting populations. Ziegler points out that studies on the community level have been rare (op cit., p. 3956).

Evolutionary Theory

The birth, growth, change and disappearance of firms are processes that take place in interaction with the ecosystem. In evolutionary theory a framework is built for describing and explaining these processes. Evolution – of which an ecosystem is its temporary manifestation – results from the operation of four generic processes: variation, selection, retention and diffusion, and the struggle over scarce resources (Campbell, 1969). Below follows brief definitions of these notions.

Variation is defined by Aldrich, (1999, p. 22) as: 'Any departure from routine or tradition is a variation, and variations may be intentional or blind.' Intentional variations are subject to some kind of decision process and attempt to generate alternatives and seek solutions to problems.

'Forces that differentially select or selectively eliminate certain types of variations generate a second essential evolutionary process: *selection*' (Aldrich, 1999, p. 26). Selection criteria are set through the operation of market forces, competitive pressures, the logic of internal organizational structuring, conformity to organizational forms, and other forces. Selection processes are crucial for organizational survival and organizations that command fewer resources from their environment, due to for example maladaptive variations in technology, are likely to disappear or decline in performance.

'Retention occurs when selected variations are preserved, duplicated, or otherwise reproduced so that the selected activities are repeated on future occasions or the selected structures appear again in future generations' (Aldrich, 1999:30). Retention processes are the mechanism by which organizations make use of routines and their 'organizational mind' (Hedberg and Wolff, 2001). The success of retention processes is a function of the pace of environmental change. On the population level, retention processes guarantee the preservation of specific competencies. For example, if one particular type of company goes bankrupt the employees can be transferred to other similar organizations in need of that competence and thus the knowledge is preserved.

Struggle is the underlying pressure for selection and variation (Aldrich, 1999, p. 32). Scarcity of resources within organizations, between organizations and between populations is the dynamic force, which fuels the evolutionary process as such.

The evolutionary approach, using the above-mentioned four principles, explains how particular forms of organizations come into existence in specific kind of environments. In practice, all four principles occur simultaneously rather than sequentially. They are linked in continuous feedback loops.

Community and Ecosystems

The term that is closest to our understanding of ecosystems is Aldrich's 'community' (Aldrich, 1999). He defines community: '... as a set of co-evolving populations linked by ties of commensalism and symbiosis ... Within a community, processes of competition and cooperation sort populations into differentiated niches, and dominant populations drive others into subordinate positions and ancillary roles' (op cit., p. 298). Aldrich describes 'the commercial community of biotechnology' and the 'commercial community of the web' in terms of *governance structures, commercial users, usage promoters, infrastructural populations and core technologies.* New populations occur when three kinds of discontinuities exist:

- shifts in social norms and values;
- changes in laws and regulations;
- technological innovations (op. cit., pp. 311–312).

Organizational ecology has not examined the processes involved in organizational startups and entrepreneurship (Aldrich, 1999, p. 46). In practice also ecologists adopt population definitions that correspond closely to industrial economists' conceptions of industries (op. cit., p. 47), looped over decades as an outcome of many different evolutionary processes.

Two additional concepts are central for the understanding of relations between populations in the ecosystem: symbiosis and commensalism. Symbiosis denotes a mutual dependence between dissimilar units, whereas commensalism means that units make similar demands on the environment. The term commensalism is often used to mean cooperation or mutualism. But, Aldrich (1999, pp. 301–302) describes commensalism in terms of populations seeking the same resources. Commensalism therefore can take on different forms, from competition to full mutualism. From an ecosystems perspective these processes do not need to be intentional, but rather an outcome of evolutionary processes between and inside organizations.

Niches, Strategies and Theory of Business

Organizational learning theories have had a too narrow focus on adaptation (Hedberg and Wolff, 2001). It seems that selection processes are more or at least equally important to the evolution of single organizations. Organizations select niches. The width or reach of a niche is defined 'as the variance of its resource utilization and can vary differently along multiple dimensions' (Ziegler, 1996, p. 3960). Organizations that have selected a wide niche are classified as *generalists* and those choosing a narrower niche are named *specialists*. Using the language of strategy these processes can be classified as *positioning*, with the difference that the (normative) strategy approach usually assumes positioning to be a rational process. The ecological approach has a different perspective; strategies are emergent and are an outcome of the constant interplay between the ecosystem and its parts; selection processes are the manifestations of these exchanges.

To understand the evolution of industries and specifically how single companies develop their theory of business the specific ecosystem has to be understood. The ecosystem plays an important role regarding the processes of variation, selection and retention in the struggle for competitive advantages. The industry perspective can differentiate from the population perspective in the following two ways:

- When we describe industries we usually refer to more homogeneous businesses using similar technologies and acting in the same kind of commercial niche.
- Industries also can be activity systems, where the parts are single interconnected companies and these companies have complementary functions in the system as a whole, leveraging solutions to specific customers.

Whereas industry populations are more homogeneous by contents, activity systems are characterized through diversity and complementary relations. Ecosystems are regional systems, local habitats that may consist of many populations of which industry populations and activity systems are parts. Governance structures and actors are part of the ecosystem.

In conclusion, ecosystems are constituted by:

- community and governance structures;
- industry populations;
- activity systems.

New industrial populations occur when three types of discontinuities exist:

- shifts in social norms and values;
- changes in laws and regulations;
- technological innovations (Aldrich op. cit., pp. 311–312).

Repositioning is triggered in a company when the management acknowledges that either discontinuity is relevant for it, or a combination of those. The global organizational eco-system containing the oil industry is currently exposed to such potential discontinuities and we would like to illustrate the concepts presented above with an example of some oil companies' process of repositioning within this system.

Repositioning the Oil Industry

For many years the petroleum sector has been criticized for its business activities. Environmental organizations, public authorities and consumers have formulated the criticism. The petroleum sector itself has been united in public relations about the problem of global warming and climate change, backed by other industrial sectors and lobbying organizations, until 1996. That year, BP (British Petroleum) left the largest lobbying organization and wanted to take an alternative position to this serious global environmental problem of climate change. Even if it was not 100 per cent scientifically proven that there was an enhanced green house effect, BP said they wanted to take a more responsible position. Shortly afterwards, the Royal Dutch/Shell Group announced a similar opinion and claimed that they were committed to sustainable development.

This was the beginning of a repositioning process in the petroleum sector, and in the two companies in particular. They wanted to be viewed as 'energy-providers' rather than 'just' oil companies and within the term 'energy' renewable energy sources were included. BP and Shell can be seen as 'early movers' in the sector and have had increasing investments and growing business areas in renewables since the start of the repositioning process; BP mainly in solar technology and Shell in solar, forestry, biomass and wind power. The two companies lead the development in the solar power area (Energimyndigheten, 1999) and do research together with large automotive corporations in hydrogen technology as well. In addition to this, the two companies have been and are involved in other activities such as implementing corporate responsibility and business ethics in the ordinary business work. Activities and reporting in the Health, Safety and Environmental (HSE) areas are increasing and developing steadily although the industry lacks common reporting standards which make comparability among the competitors more difficult (SustainAbility et al., 1999). Both BP and Shell think that the importance of renewable energy will increase in the future and the costs of producing electricity from solar power, biomass and wind are steadily falling, although a shift in the global energy system, from fossil fuels to renewables, will take a long time. This is something that both companies also admit. In the meantime the more 'environmentally friendly' natural gas (in comparison with coal and oil) will play a central role and its share in the companies' energy portfolios is increasing steadily. As a part of this process of modifying its image and business model, BP has changed the company name: BP, the former British Petroleum has been transferred into 'Beyond Petroleum'!

Some scholars such as Cowell et al. (1999) mean that the oil industry has little choice but to claim their commitment to sustainability, and that they enter the debate with initial contempt. Other scholars have a different opinion or a more positive standpoint; that renewable energy can be profitable, it is only a question of when (Faucheux and Nicolaï, 1998; Rannels, 2000; Sayigh, 1999; Stuebi, 1999). The latter view gains support through the fact that, apart from BP and Shell, other oil companies involved in the development of renewables are: Agip, Aral, Exxon, Mobil (Mobil Solar), Neste (NAPS Finland), Statoil, Texaco and Total (Total Energie France) (Energimyndigheten, 1999; Oliver and Jackson, 1999).

The petroleum and energy business have an important role in making the world a more sustainable place to live in. It is therefore welcomed and very interesting when this sector seems to have taken at least a small step further towards more sustainable sources of energy. But at the same time, this change or mind broadening raises some questions around the multinational corporations in general and the oil industry in particular. Accidents such as the oil spill from Exxon Valdez in Alaska 1989, indications about Shell and its supposed interference in Nigeria's internal politics (true or not, the rumors have probably damaged the company and its reputation), the dumping of Shell's Brent Spar[1] in 1995 etc. have given the business a bad reputation. The latter incident resulted in costly damage in both European sales and reputation for Shell (Grolin, 1998; Neale, 1997). So, the business certainly could do with a higher reputation. Is all this talk about sustainability and the initial repositioning a groundless discussion, or is it a result of discontinuities in norms, technologies and regulations affecting in the oil industries' organizational eco-system? Here we try to find evidence of changes at a low level, in the activity systems of two selected companies, which may point to a repositioning in this industry.

BP – an Example for Repositioning a Business Model in a Changing Ecosystem

Our study of BP contained two elements: the emergence and presentations of the 'new theory-in-business' as well as a first description of the changing components of its activity systems.

In 1996 a process started in which the management and owners of the company began to reevaluate the assumptions and foundations of their businesses. The analysis of BP's managements policy speeches during 1997 (Beausang, 2000, p. 41) shows that among the most used words that year were 'environment', 'oil' and 'green house'. In 1998 'green house' was replaced by 'gas' as one of the three most frequently used terms. In 1999 and 2000 the process of formulating a new position in the market found new ground:

---

[1]Brent Spar was an offshore storage and tanker loading buoy owned by Shell UK Exploration and Production and was situated in the Brent Field in the North Sea. The buoy was dumped in the sea in 1995 and caused many negative reactions from organizations and the public.

We are totally committed to sustainable energy solutions, not as public relations froth …
But as the essential membership ticket to the club of responsible corporate citizens in the
new Millennium.
(6 January 2000)

Learning about the challenges in different parts of the world and learning about the
potential responses and our role in creating environmental progress.
(27 April 2000)

There is also a changing concern expressed regarding the increased green house
effect.

The gathering evidence of a fundamental change to the climate caused by human
activity.
(26 April 2000)

The latest authoritative scientific reports on climate change make clear – in the most
careful, rigorous language – that indications of a human effect on the climate are
mounting.
(26 April 2000)

In other words a company's search for a new rhetoric (=theory-of-business) is not
trivial at all. It reflects an internal learning process, maybe even extensive internal
power struggles, and a struggle and obligation to make the company survive. The
oil industry has an extremely long-term investment cycle (Zaring, 1999). The
allocation of investments therefore is a fundamental restriction to the speed in
which activity systems can be changed. The new theory-of-business induces internal
learning, a change process during which the rationale for the total business is
questioned and the new business model that emerges re-evaluates the internal
allocation of resources to defined business areas.

In 1996 a new gas company called Alliance Gas was formed, with to increase
the market share in natural gas. In 1997 BP formed a strategic alliance with AO
Sidanco to bring management competence to the Russian energy market (a 570
USD million investment). In 1998 BP merged with Amoco, their purpose being to
increase gas market share. Each year since 1994 BP invests 30 million USD in
solar energy and in 1998 BP opened a new solar plant in California and has today
such plants in the US, Spain, Australia and India. Amongst many other activities, in
late 2000 BP announced that the group intends to invest USD 500 million in lower
carbon energy markets i.e. photovoltaics and hydrogen technology. The process of
repositioning BP as an energy provider thus started with the new theory of business,
influenced the business model and has started to change the structure of the
business system, its components, its resource base and not the least, its performance
criteria.

Why Then the Use of Organizational Ecology as a Theoretical Frame of Reference?

Organizations reside in ecosystems and are part of organization fields. DiMaggio and Powell (1983) described that '... organizations, in the aggregate, constitute a recognized area of institutional life: key suppliers, resources and product customers, regulatory agencies, and other organizations that provide similar services or products' (p. 148). Institutionalization in the field constitutes the mechanism that BP and its competitors adhere to. They even organized a common ground for expressing their rejecting of the 'green house threat'. Selznick (1957) called to institutionalize 'to infuse with value beyond the technical requirements of the task at hand' (p. 17). Scott (1987) defines the creation of institutions as a process where actions are repeated and come to mean the same thing to organizations within a field. This was what constituted the industries' compact resistance to concede in public its contribution to the green house threat. When BP and Shell left this institutionalized sphere the global coalition started loosing its institutional power.

Organizations find niches in order to defend themselves against competition. The habitat, as we call it, in which these organizations act has a non-trivial impact on the choice of niches. A company's business theories, business models and business systems are an outcome of the sensemaking of the niche within the company. The reshaping of a new institutionalized agenda for the oil industry has perhaps also elements of mutualistic cooperative relations. BP cooperates with many players in its industry. BP and Shell left the Global Coalition almost simultaneously – a coincidence? Both companies are European companies and the question is whether the European inheritance and habitat had an influence on how these companies have reacted to stakeholder demands. Clearly, the European consumers are more inclined to relate to environmental issues than American consumers. Clearly, the European financial market has a much clearer focus on discriminating oil companies (in investment funds and related financial products). This all has had effects, although the causal relations are difficult to demonstrate.

BP also is an interesting example of selection, rather than adaptation. The company chose to depart from routine and tradition. It was an intentional choice and generated new solutions and structures to problems. The selection was conducted in an institutional setting with a strong coherence amongst competitors to structure the market in a traditional way. BP broke with this tradition and selected a new niche in the ecosystem. Doing this the company had to learn new technologies, new markets and find new coalitions. Clearly, BP envisioned a coming scarcity of oil resources, but also a scarcity of future support amongst its shareholders and stakeholders.

BP chose a new niche, and in so doing repositioned the company. The old niche 'oil producer' gradually changes towards 'energy provider' as the components of the company are rearranged, performance criteria are reevaluated and resources are reallocated. This emergent strategy will be learned in the constant interplay with the company's ecosystem. The process of selection will step-by-step become retention processes, which will take decades to settle.

Sustainable Repositioning as an Evolutionary Process

This contribution represents an ongoing research agenda for us. More than a decade of environmental management research has strengthened our opinion that management research is vital in the shaping of a more sustainable globe. The understanding of how business organizations contribute to this change, the broader conceptualization of the institutional ecosystem in which these changes occur and our normative stance, that those companies that dare to take the risk of selecting new niches have to be supported, has grown stronger. The organizational theory frame of reference that we adapt opens up for theoretical mutualism with other disciplines, such as the sociology of law, sociology, environmental economics and psychology.

Our own research agenda has set out to further develop the repositioning concept in the ecosystem perspective. Much further empirical, longitudinal studies are asked for. The challenge ahead is both theoretical and practical. The practical part of the challenge is to support those business companies that dare to take the step to challenge their own theory-of-business, against the established traditions in their industry. These standard setters deserve our support, criticism and our respect!

References

Aldrich, H. E., (1999), *Organizations Evolving*, London: SAGE Publications.

Beausang, P., (2000), *Sustainable repositioning? A pilot study of two oil companies*, Göteborg: Gothenburg Research Institute.

Bennulf, M., (1994), *Miljöopinionen i Sverige*, Lund: Dialogos.

Berle, A. A., and Means, G. C., (1995), *The Modern Corporation and Private Property*, New York: Macmillan. Original edition 1932.

Black, S. W., et al. (1998), *Competition and Convergence in Financial Markets*, Elsevier: Amsterdam.

BP (1996), *HSE Facts*.

BP (1997a), *HSE Facts*.

BP (1997b), *Social Report*.

BP (1998a), *General Business Policies*.

BP (1998b), *BP Environmental and Social Report*.

BP (2000), *China and Renewable Energy*.

BP Amoco (1999a), *Alive*.

BP Amoco (1999b), *Strategy Presentation 1999*, presentation made by the CEO of BP Amoco, Sir John Browne.

BP Amoco (2000a), *Statistical review of world energy June 2000*.

BP Amoco (2000b), *Development and implementation of a process to audit BP Amoco's greenhouse gas emissions – Audit overview February 2000*.

Bragd, A., (1998), *Planerarna, fixarna och bläckfisken – en organisations erfarenheter av miljöanpassade produkter* (The planners, the thinkers and the octopus – an organisation's experience of green products), Göteborg, Göteborgs universitet , Licentiate thesis.

Campbell, D. T., (1969) 'Variation and Selective Retention in Socio-Cultural Evolution', *General Systems*, 14: 69–85.

Carlsson Reich, M., Wolff, R., et. al. (2001), *Ethical Investments – Towards a Sound Theory and Screening Methodology*, Stockholm: IVL Swedish Environmental Research Institute report.

Copeland, T. E., et al. (1994), *Valuation: measuring and managing the value of companies*, New York: Wiley.

Cowell, S., Wehrmeyer, W., Argust, P., and Robertson, G., (1999), 'Sustainability and the primary extraction industries: theories and practices', *Resources Policy*, issue 25, Pergamon.

Czarniawska, B., and Joerges, B., (1996), *Travels of Ideas*, in B. Czarniawska, and G. Sevón (eds) *Translating Organizational Change*, Berlin: de Gruyter.

Daly, H., (1992) 'Allocation, distribution, and scale: towards an economics that is efficient, just, and sustainable', *Ecological Economics* 6:185–193.

DiMaggio, P. J., Powell, W. W., (eds) (1983) *The New Institutionalism in Organizational Analysis*, Chicago: University of Chicago Press.

DiMaggio, P. J., and Powell, W. W., (1991), 'Introduction', in P. P. DiMaggio and W. W. Powell, (eds) *The new institutionalism in organizational analysis*, Chicago: The University of Chicago Press.

Dobers, P., and Wolff, R., (2000), 'Competing with "soft" issues – from managing the environment to sustainable business strategies', in *Business Strategy and the Environment*, pp. 143–150.

Drucker, P., (1994), 'The Theory of Business', *Harvard Business Review*, 72/5: 95–104.

Energimyndigheten (1999), *Förnyelsebara energikällor*, ER:1999, Eskilstuna.

Faucheux, S., and Nicolaï, I., (eds) (1998), *Sustainability and firms: technological change and the changing regulatory environment*, Cheltenham : Edward Elgar.

Friedman, M., (1962), *Capitalism and Freedom*. Chicago: The University of Chicago Press.

Friedman, M., (1970), 'A Friedman Doctrine – the Social Responsibility of Business is to Increase its Profits', *The New York Times Magazine*, September, 13, 32–33 and 123–125.

Friedman, M., and Friedman, R., (1980), *Free to Choose*, New York: Avon Books.

Grolin, J., (1998), 'Corporate legitimacy in risk society: the case of Brent Spar', *Business Strategy and the Environment*, issue 7, pp. 213–222, John Wiley & Sons.

Hamel G., (2000), *Leading the revolution*, Boston, Mass: Harvard Business School.

Hannan, M. T. and Carroll, G. R., (1992), *Dynamics of Organizational Populations: Density, Legitimation, and Competition*, New York: Oxford University Press.

Hannan, M. T., and Freeman, J. H., (1989), *Organizational Ecology*, Cambridge, MA: Harvard University Press.

Hawley, A., (1950), *Human Ecology*, New York: Ronald.

Hedberg, B., & Wolff, R., (2001), 'Organizing, Learning and Strategizing: From Construction to Discovery', in M. Dierkes et al., *Handbook of Organizational Learning*, Oxford University Press, pp. 535–556.

Jaffe, A. B., Peterson, P. R., and Stavins, R. N., (1995), 'Environmental Regulation and the Competitiveness of U.S. Manufacturing: What Does the Evidence Tell Us?', *Journal of Economic Literature*, Vol. XXXIII, March, pp. 132–163.

Jensen, M. C., and Meckling, W. H., (1976), 'Theory of the Firm: managerial behaviour, agency costs, ownership structure', *Journal of Financial Economics* 3:305–360.

March, J. G., (1994), *Three Lectures on Efficiency and Adaptation in Organizations.* Helsingfors: Swedish School of Economics and Business Administration.

Mintzberg, H., Ahlstrand, B., Lampel, J., (1998), *Strategy Safari – a guided tour through the wilds of strategic management*, Harlow: Financial Times Prentice Hall.

Neale, A., (1997), 'Organisational learning in contested environments: lessons from Brent Spar', in *Business Strategy and the Environment*, Vol. 6, pp. 93–103.

Oliver, M., and Jackson, T., (1999) *The market for solar photovoltaics*, Energy policy issue 27, Elsevier.

Pfeffer, J., (1993), 'Barriers to the Advance of Organizational Science: Paradigm Development as a Dependent Variable', *Academy of Management Review*, 18, 4 (October): 590–620.

Porter, M. E., (1990), *The competitive advantage of nations*, New York, Free Press, 1990.

Porter, M. E., (1991), 'America's Green Strategy', *Scientific America*, April.

Porter, M. E., and van der Linde, C., (1995), 'Green and Competitive: Ending the Stalemate', *Harvard Business Review*, Sept.–Oct., pp. 120–134.

Rannels, J., (2000), 'Advancements in the United States Photovoltaic Program', *Renewable Energy* issue 19, Pergamon.

Sayigh, A., (1999), *Renewable energy – the way forward*, Applied energy issue 64, Elsevier.

Scott, R. W., (1987), 'The Adolescence of Institutional Theory', *Administrative Science Quarterly*, 32:493–511.

Selznick, P., (1957), *Leadership in Administration*, New York: Harper & Row.

Stuebi, R. T., (1999), 'New Power Technologies: Too important and interesting to ignore', November issue, Elsevier.

SustainAbility, UNEP and Engaging Stake Holders (1999), *The oil sector report – a review of environmental disclosure in the oil industry*, Beacon Press. London.

Symreng, T., Wolff, R., and Zaring, O., (1999). *Miljöfondmanagement – Risker och framtida möjligheter*, Uppdragsrapport, 6:e Ap-fonden och Skandia.

The World Commission on Environment and Development (1987) *Our Common Future*, Oxford: Oxford University Press.

Thurow Lester, C., (1999), *Building Wealth: The New Rules for Individuals, Companies, and Nations in a Knowledge-Based Economy*, New York: Harper Collins.

Vickers, G., (1995), *The Art of Judgement. A Study of Policy Making*, Sage, London.

Wolff, R., and Zaring, O., (1999), *Indicators for Sustainable Development*, Gothenburg Research Institute, Göteborg.

Zaring, O., (1999), *Investment decisions and the environment*, BAS, Göteborg.

Ziegler, R., (1996), 'Organizational populations', in: M. Warner (ed.) *International Encyclopedia of Business and Management*, London: Routledge.

Legal and Governing Strategies –
Towards a Law of Sustainable Development

Håkan Hydén and Minna Gillberg

If you don't change course, you will end up where you are headed

Chinese proverb

Introduction

A contaminated river, air pollution from car exhausts and damage to the ozone layer are examples of environmental problems which must be addressed in two respects. One aspect concerns the natural environment in itself, i.e. how we define and analyse what is taking place within the natural systems. This is a task that our society has single-handedly entrusted to the natural scientists. The other aspect of the issue concerns the dilemma of actually solving the environmental problem. In this respect we are talking about the different technical measures that which can be applied. These technical measures rely on data provided by natural science, and try in one way or another to influence the processes taking place in nature. However, there is an obvious dimension that exists as a hindering filter between the environmental problem and its solution, namely, the human link and condition. Someone has to change behaviour, take measures and assume responsibility. This is where the societal context at large and, in particular, its strategic regulation mechanisms (governance), come into the picture.

The focal point for this anthology lies on how we can influence, and alter, human behaviour so that it becomes more environmentally friendly and eventually, sustainable. This article wants to elaborate on what the legal, and normative, dimension can contribute when it comes to tools for problem solving and creating new regulation strategies, particularly as regards industrial practice. The legal instrument is in a metaphorical way parallel to the technical tools that are geared to the practical aspects of the environmental problems. Law plays a corresponding role in relation to human and organizational behaviour. It is one tool to be used for the purpose of changing or channelling societal behaviour. Law is inherently normative. But as little as one can disregard natural scientific knowledge when deciding what technical measures should be taken, one cannot leave out the societal aspects of normative structures in society when designing legal instruments and regulation strategies.

In this article we will start by discussing how environmental problems, following in the track of industrialisation, were perceived and addressed by Swedish society during the twentieth century. The way environmental issues have been tackled by the legal system may be said to reflect how society has viewed and prioritized the natural environment. (It should be noted here that the conclusions we draw, based on the Swedish experience, are in the main applicable to a wider context of continental European law when analysing how such legal systems have addressed environmental problems.) At first sight the Swedish approach to environmental regulation constitutes a classical command and control model. However, at close range, the Swedish model constitutes a compromising, consensus-oriented, decision-making procedure rather than a clear-cut command and control approach.[1] This model was confirmed by the Swedish Environmental Protection Act of 1969, and in several ways reinforced by the new Environmental Code of 1999.

In the following we will address the fundamental weakness of the command and control paradigm and then continue to discuss the emergence of a new regulation paradigm. The roots and the force of this new paradigm are to be found outside the traditional administrative and politically based regulation system, namely, in the interplay between the so-called market and the grass-roots activities of the civil society. This new regulation paradigm builds upon two prerequisites. Firstly, the shift from negative to positive economic and value-based connotations as regards the environment, and secondly, the shift in a societal regulatory perspective from 'command and control' towards 'demand and sustainable development'. Finally, we conclude by defining the embryo of a (governing) legal regulation strategy that normatively strives towards the creation of a sustainable society. Furthermore, in this context, we outline a concrete implementation tool for such a strategy, a Sustainable Management System.

---

[1]Minna Gillberg, *From Green Image to Green Practice – Normative action and self-regulation*, pp. 31–47, Lund Studies in Sociology of Law 6, Lund University, 1999.

The Problem Formation of Environmental Regulation

Today, facing a potential environmental global disaster, it is not difficult to draw the conclusion that there seems to be a great and inherent contradiction between the present legal regulation system and the concept of sustainable development. Legal rules are a product of the political system. Politics as such is inherently not sustainable. Politicians come and go, as do the ideas guiding their politics. As long as the political system and the infrastructure of our society, like the tax system, are not based on an economic rationale which integrates a long-term sustainable perspective, legal regulation attempts in the field of environment cannot be successful, i.e. sustainable. We are faced with a similar problem when it comes to integrating the concept of sustainability into the legal system, which, to put it bluntly, does not constitute one unity. The legal system is shaped by different forms of rules directed towards separate parts, or functions, of society. Civil law, penal law, procedural law and administrative law constitute dissimilar legal branches. Civil law and penal law have traditions that go back thousands of years, sprung out of our socio-cultural and economic systems. These rules establish limits – a code of conduct – for social and economic behaviour and provide the instruments for co-operation between individuals and organizations. This part of the legal system is formally completely indifferent to the environment since the clear-cut task is to provide instruments for regulating human interaction. The rationale of the legal system is to create stability and foresee-ability. This implies that if one has the ambition to alter social or economic behaviour into a more sustainable mode, the prevailing legal system intrinsically has to alter its *modus operandi*.[2]

The occurrence of large-scale environmental problems, with a global impact, is in the main coupled with the rise of industrialization. Hence, environmental legislation is a modern phenomenon on the legal scene. The increasing use of energy during the industrial era has led to a situation where the societal cycles of production, distribution and consumption of goods and services are spinning faster and faster without considering the impact on the ecological cycles. Thus, it is not surprising that, when environmental legislation was considered for the first time in Sweden, in 1909, the intention of parliament was to protect the emerging industry against legal claims for prohibition and damages rather than to protect the environment. The objective of the law was simply to facilitate, in an organized and rational form, the exploitation of the ambient environment through a licensing system.[3]

---

[2]Perhaps we will eventually see the emergence of such a necessary shift in political, economic as well as legal rationale through the new articles inserted on sustainable development into the Amsterdam Treaty (see the preamble and Article 2 of the Treaty on the European Union, as well as Articles 2 and 6 of the European Community).

[3]See Hydén, H. (1998), p. 29.

In retrospect, the environmental disturbances caused by industrial activities manifested themselves within the existing legal practice. The environmental disputes had to be versed in terms of general legal principles, which reflected society's view of the right to exploit nature. Hence, the articulation of environmental conflicts brought to court was restricted to the framework, concepts and technical argumentation of the predominant legal jurisprudence. In other words, the legal system lacked adequate tools, and was therefore not prepared to address these conflicts, except through the concept of property. However, the process of industrialization unravelled a contradiction within the concept of property itself. The contradiction manifested itself through the 'right to do' something and the 'right to exclude' others from doing; i.e. the contradiction between the right of disposition over one's own property and the right to enjoy protection against disturbance.

The 'right to do' entails a right to freely exploit the property you own, while the 'right to exclude' entitles call for actions against a proprietor who, while exploiting his property, causes negative effects on another person's property. In the perspective of market economy, property rights, in terms of rights to disposition, are functional, while the 'right to exclude' can be viewed as a privilege deriving from the feudal world of ideas. Given this, it is appropriate to ask how politics and jurisprudence have addressed the inherent contradiction of property rights, when it comes to environmental problems. When we look at how legal practice has approached the matter during the twentieth century, it is evident that two different ways of embracing environmental disturbances have emerged. One path, 'the right to do', is made up of civil law and the claim for damages, while the other path, 'the right to exclude', consists of nuisance and the call for prohibition.[4]

The civil law strategy, as a means of solving the conflict between the two sides of property claims, is characterized by compensation for damage, paid by the exploiting proprietor to the proprietor who suffers damage, corresponding to the value of the damage. The objective of this strategy is merely to reach a compromise, through economic compensation, that can be accepted by the proprietor whose right is being violated. In contrast, the objective of the nuisance strategy is to restore the freedom of the infringed proprietor. In this case the solution is to prohibit the exploiting proprietor from continuing the activity, or to put an injunction on the exploiter to implement precautionary measures so as to put an end to the damage.

It did not take a long time before the principles behind the institute of nuisance

---

[4]This is dealt with in Håkan Hydén, (diss.) *Rättens samhälleliga funktioner*, (*The societal functions of law*), chapter 8, Lund 1978 and in the article 'Miljöetik och miljörätt' (Environmental Ethics and Environmental Law) in Håkan Hydén, *Rättssociologiska perspektiv på hållbar utveckling*, (*Sociology of law perspectives on Sustainable Development*), Sociology of Law Research Reports, 1998:1. For a brief version in English, see Håkan Hydén, 'Towards a Postinterventional Environmental Law', in *Jahreschrift für Rechtspolitik* 1991, pp. 245–259.

were questioned. Industrialization required foreseeability, operationalized through political control over natural resources and production conditions. The economic system called for political action and ultimately for legal innovation. Hence, the bill for the Code of Land Laws of 1909 proposed an industrial licensing system. The Bill stated that it was necessary to avoid a development whereby legal practice would paralyse industry and hamper growth. Therefore, landowners and others had to tolerate a certain amount of nuisance from another landowner's activities. In order to ensure economic growth, Swedish society found it necessary to achieve consensus and compromises between conflicting interests, by creating an institute of intervening norms in the field of environmental law.

The Creation of a Compromise and Control Paradigm

It took 60 years before the initial proposal on a system of licensing industrial activities (the bill for the Land Laws of 1909) was realized through the 1969 Environmental Protection Act (EPA, SFS 1969:387). The act was an attempt to create a comprehensive, all-embracing piece of legislation, connecting conservation, species preservation, chemicals, toxins, energy, transport and development planning. In practice, it turned out to be somewhat incomplete and uncoordinated, not to say dysfunctional. The EPA constitutes an illustrative example of how the well-known Swedish consensus model has been embodied in Swedish legislation. It is a legislative technique which is referred to as an 'open-ended' framework law, containing the goals, broad guidelines and principles which are to be operationalized through detailed regulations issued by government, authorities or by the county administrative boards.

The EPA was enforced within the administrative structure by the National Licensing Board of Environmental Protection (Koncessionsnämnden för miljöskydd). The Licensing Board, set up in 1969, was under the jurisdiction of the Swedish Government (acting as the highest administrative court of appeal) and issued permits and directives on emission levels for 'large-scale' environmentally hazardous activities. The County Administrative Board granted permits for less significant environmentally hazardous activities. Since it was an administrative procedure, the national government acted as the highest administrative court of appeal – a somewhat problematic role as regards the security of 'life and property' in Article 6 § 1 of the European Convention on Human Rights.[5] The procedure

---

[5]The European Court of Human Rights found in the Case of Zander vs Sweden (45/1992/ 390/468) that the proceedings before the Licensing Board under the 1969 Environmental Protection Act did not meet the demands regarding rule of law, in relationship to property and civil rights, as stated in article 6 § 1.

under the Licensing Board has been viewed by the Swedish state as the most important legal instrument for environmental regulation and protection in Sweden. This justifies a closer examination of the mechanisms in the system when, in hindsight, it comes to identifying and explaining unsuccessful attempts of this regulation strategy in safeguarding the environment.

As mentioned earlier, the system of a Licensing Board was first proposed at the beginning of the twentieth century. At that time, the main purpose behind the proposal was to protect industry from compensation claims from parties suffering damage from industrial activity. The underlying idea was to solve the problem in advance by achieving a consensual compromise. When the Licensing Board system was introduced and implemented in 1969, the final outcome did not deviate too far from the 60-year-old original proposal.[6] The underlying rationale of the EPA was that environmental interests were to be weighed against employment issues, public welfare and economic factors. The Swedish consensus model, its legislation and policy making defined environmental problems as a question of what industry could bear in terms of the costs of precautionary measures. At that time, it was not primarily a question of how much pollution and ecological disturbance the environment could tolerate. The decision-making process of the Licensing Board solved an environmental issue purely as a political and economic problem, not as an environmental (ecological) problem.[7]

The provisions in the EPA only pointed out certain factors to be taken into consideration in the decision making, but did not state anything about how these factors were to be weighed against each other. This evaluation was left to the Board, and the entire procedure could be described as a process of providing industry with guiding advice, based on compromises instead of command.[8] In addition, the law demanded that the concept of best available techniques should be applied. However, it was up to the authorities to prove that a certain technique existed, and also that it would work at the specific plant applying

[6]Håkan Hydén, *Rättens samhälleliga funktion (The Societal Functions of Law)*, 1978, chapter 8.

[7]Minna Gillberg, *From Green Image to Green Practice – Normative action and self-regulation*, pp. 31–47, Lund Studies in Sociology of Law 6, Lund University, 1999.

[8]Since the early 1970s, governments throughout the industrialized world have responded to the rise of environmental degradation and industrial pollution with a myriad of environmental policies. The dominant government response, however, has according to Neil Gunningham et al., been the application of direct or command and control designed to prohibit or restrict environmentally harmful activities, Neil Gunningham and Peter Grabosky, *Smart Regulation. Designing Environmental Policy*, Oxford (1998). For an overview, see R. Kagan, 'Regulatory Enforcement', in D. Rosenbloom and R. Schwartz (eds), *Handbook of Regulation and Administrative Law*, New York, (1994). The Swedish model, though, is a consensus model which means that the compromises have to be worked out in praxis.

for a permit. A consequence of this rationale was that it did not create incentives for the polluting party to initiate technical improvements in order to meet the demands of the environment. Another guiding principle was that the measures called for by the authorities, must be economically feasible and reasonable, i.e. the authorities could not impose other measures than those lying within the economic range of the enterprise. In practice, this created a situation where, if an efficient technique existed, authorities could not impose these measures if the company argued that they were not economically feasible. The logic behind the regulation was systemic, it was first and foremost a question of maintaining a system that upheld neutrality in competition.

Another serious issue with the EPA was the fact that the proceedings of the Licensing Board were not clearly regulated in the EPA. For instance, there were no formal rules of procedure for the hearings before the Board. And more alarming still, witnesses did not testify under oath and cross-examination was not allowed. This created a situation where companies applying for permits, in several cases, lied openly before the Board about their emission levels and use of chemicals.[9] Furthermore, in principle, once the Licensing Board had issued a permit and it had gained legal force, it remained in effect for all time. The Act did in fact state that permits may be re-examined ten years after issue, even earlier if unexpected problems arise subsequent to the permit being issued or if the environmental situation can be improved through the use of better technology or industrial processes. However, the permit holder and the Swedish National Environmental Protection Agency exclusively had the right to initiate a reconsideration of the permit before the Licensing Board. The same rules applied when the County Administrative Board had issued a permit. Since the Environmental Protection Act came into force in 1969, the Board has issued more than 6,000 permits. The Agency has so far exercised its right to initiate a reconsideration of the permit in approximately 50 cases.[10] There is no doubt that environmentally friendly technology has developed considerably in the 30 years that the National Environmental Protection

---

[9]The companies would e.g. claim that their emissions were much higher than they were in reality, and when the Licensing Board then imposed, as they believed, strict restrictions (for instance a 30 per cent emission reduction), they did in fact only meet the actual emission level. In some cases the Licensing Board unintentionally even provided companies with a future license (based on false data) to pollute more than they did for the time being. See as an example, documents from Volvos proceedings before the Licensing Board, 2 March 1990. And, articles in Östra Småland 8 and 10 March 1990, Idag/GT Kvällsposten 19 March 1990.

[10]To be noted is the fact that if an industry wanted to increase its production, a new application had to be submitted to the Licensing Board (since the new Environmental Code entered into force applications must be submitted to the Environmental Court). This gave the Board a theoretical possibility to put more stringent demands as regards pollution levels.

Agency has acted as an enforcer. This new technology must be applicable to a lot more than the 50 cases of hazardous industrial activities where the Agency has reconsidered and issued new permits. As a result, numerous environmentally hazardous activities are operating today on the basis of old permits, outdated techniques and causing unnecessary environmental damage. This also had the legal implications that if a person suffered damage from an environmentally hazardous activity performed by a permit holder, she or he had/has no legal resource to stop the activity, or to force the permit holder to limit the extent of the damage. It was (and is) only possible to lodge a claim for compensation. Thus, the permit protects the offending permit holder in perpetuity from attempt to seek redress through the courts. Furthermore, it provides impunity for actions otherwise regarded as unlawful.[11] It should be noted here that the problematic structure of the licensing procedure, its decision-making body, has been exported to the new Environmental Code (SFS 1998:808). The only substantial difference to speak of is that the Licensing Board is relabelled and referred to as the Environmental Court.

The intention of the Environmental Code of 1999 was again to create an all-embracing code. However, important legislation regulating forestry, agriculture, infrastructure, transportation and energy issues were left outside the Environmental Code. In certain cases (listed in chapter 17 § 1), the decision-making procedure sidesteps the environmental court and gives authority to the government to grant permission for activities regarded as being of national interest, despite the fact that they can have a negative impact on the environment. Here, we may mention that the importance of the activity for employment qualifies as an example of a special circumstance deemed as being of national interest.

The Environmental Code has been sharply criticized for lacking new innovative approaches. The reversed burden of proof, found in chapter 2, section 1, is however an interesting step forward in which the legal subject is liable to prove that the general rules of consideration of the Environmental Code are complied with. Furthermore, among the environmental principles (such as best available technologies, the precautionary principle and the principle of substitution), the precautionary principle is defined in the Environmental Code as a fundamental rule, whereby anyone who pursues an activity shall 'take all precautions that are necessary in order to prevent, hinder or combat damage or detriment to human health or the environment'. However, there is a 'hitch', an impediment hiding behind this far-reaching statement. While providing these progressive general rules of consideration in favour of the environment, the Environmental Code explains in chapter 2, section 7, that these rules apply only to the extent of economic feasibility, which is based upon the financial evaluation provided by the applying party.

---

[11]See chapter 29, section 1, in the new Swedish Environmental Code, SFS 1998:808.

On several occasions the Environmental Code presents cases of contradictions and systematic legal 'double standard'. According to chapter 1, section 1, the Environmental Code shall be applied in such a way as to ensure that human health and the environment are protected against damage and detriment, whether caused by pollutants or other impacts. At the same time, chapter 2, section 9, states that even if it is clear that an activity or measure is likely to cause significant damage or be detrimental to human health or the environment, the activity or the measure may be undertaken given certain circumstances. In addition, we have also seen that once a permit is granted, industry is secured to continue its activity without interference. Thus it is relevant to underline that the Environmental Code does not grant human rights to be operationalized by individuals. On the contrary, we may conclude that the Environmental Code provides systemic rights for corporations and industry to exploit the ambient environment.

To sum up the discussion, the Swedish licensing system may be regarded as a social contract between the state and enterprises, not between man and nature.[12] It is up to our public authorities to negotiate and conclude deals with the enterprises. In this procedure, environmental issues are looked upon as negotiable and tradable goods, and conceptualized in terms of technical and economic problems. The outcome is that the state-governed attempt at regulation loses in terms of value and normative statements. The fundamental 'problem' lies in the fact that is the applicant enterprise that determines the foundation for the decision by the authority. The authority, or the state, cannot interfere with the crucial issues of what is going to be produced in the factory and how. Authorities can only place limited restrictions according to the principles of best available technology (BAT), which are to be applied in accordance with what are considered to be feasible economic conditions.

As we discussed initially, the type of law where the environment is 'protected' by a license to exploit, derives from the initial political considerations at the beginning of the twentieth century, when the argument was the need for protection of industry against legal claims and prohibitions in order to ensure growth. The underlying rationale of the environmental legislation is determined by an economic rationale where the question to be answered by the law is the following: How much can an enterprise bear in terms of costs for BAT and other precautionary measures? An ecological approach would instead ask: How much can the ecosystem bear in terms of pollution and disturbance? The reason why this last question has to stand back for the first one, the economic approach, can be related to the history of environmental legislation and how the legal system did not have adequate tools to

---

[12]Sverker Sörlin, *Naturkontraktet. Om naturumgängets idéhistoria*, Stockholm, (1991), explains the difference.

address these issues in any other form than through the concept of property and the 'right to do'. The intervening norms have only extended these considerations of regulation from the individual to a larger collective, the state and our society. In one hundred years we have not travelled far. Jurisprudence today is still preoccupied with the question raised at the beginning of the twentieth century, namely, what restrictions/limitations it is reasonable to impose on economic growth/trade and the exploitation of the ambient environment. With this economic rationale as the basis for environmental legislation, important elements in legal regulation, such as morals and ethics, guilt and responsibility, fall outside the legal reasoning. Sustainable development requires that outspoken value-based rationale is also articulated within the legal system.[13] In light of the discussion above, and given the state of the environment, it is evident that the 'compromise and control model' has proved itself to be insufficient when it comes to safeguarding human life, the ecosystems and our natural resources.[14]

The Weaknesses of the Compromise and Control Paradigm

Environmental governance and regulation require knowledge and legitimacy, and in both these respects the compromise and control model faces problems in the field of environment. In order to regulate something one must have knowledge about what is going to be regulated. In the field of environment two problems manifest themselves in this respect. One concerns the interdisciplinary character of environmental problems. These have been thematized at a late stage compared to the development of science. Environmental knowledge is not structured according to the problems.[15] No science can claim to deal with environmental problems in their entirety. The knowledge is divided into different disciplines, each one without

---

[13]Minna Gillberg, *From Green Image to Green Practice – Normative action and self-regulation*, Lund Studies in Sociology of Law 6, Lund University, 1999.
[14]See Rena I. Steinzor, 'Reinventing Environmental Regulation: The Dangerous Journey From Command to Self-Control', in *Harvard Environmental Law Review* 1998, pp. 103–201. One weakness with this article is, though, that the author is focused on political solutions. 'EPA and Congress must develop better incentives for industry and public interest groups to participate in such activities' (p. 200) is a statement of that kind, while we in our article below will argue for engagement direct from the citizens as necessary for a change in a more environmental friendly way. If that does not appear, there is not much, in our opinion, that EPA and Congress can do.
[15]Cf. Håkan Hydén, 'Hur kunskapslandskapet kan förmås att matcha jordmanteln', in Tuija Hilding Rydevik, *Samspelet Mark-Vatten-Luft. Fysisk planering för att nå samhälleliga mål*, FRN Rapport 1997:1, pp. 27–49.

contact with the other.[16] Environmental science, thus, has an incomplete and fragmented understanding of the problem.

Human behaviour can be understood and interpreted in terms of norms which belong to or are created within different systems of action, such as: the socio-cultural system, the economic system, the political and the administrative system. To analyse why people act as they do in relation to the natural environment, one simply tries to identify the norms guiding their actions in the more or less specific situation. The motives of human behaviour, on an aggregated or collective scale, where incentives formed by structures of systems play a crucial role, are usually left out of epistemic science. Both natural and social sciences have so far focused on the study of observable facts and generally tended to draw a sharp line of demarcation between values and facts. Values have obviously played a greater role in legal science, moral philosophy and theology. It is primarily only within psychology that the understanding of the underlying motive for human action has been taken into account as an explanatory factor for individual behaviour. Therefore, to put it briefly, using the concept of norms as an explanatory factor and level within environmental socio-legal science, as in governance, makes it possible to integrate the essential perspectives of structures with those of actors, and their actions.[17]

A second problem in generating knowledge in this field is related to the special character of cause-and-effect relations. These are drawn out in time and space in such a way that makes it hard to create correlations between human behaviour and environmental effects. Effects of toxic substances, for instance, will usually not be detected until long after exposure, and some carcinogenic effects might not appear until later generations.[18] The environmental effects are also drawn out in space. As an example we may refer to long-range transported air pollution or water pollution. Adding to the time and space problem when it comes to identifying cause and effect in the field of environment, we have the so-called synergistic or combined effects. This means that substances react with each other in a way that is very hard to identify or predict. One chemical substance after emission might react with another chemical substance and thereby create unforeseen consequences. This is nothing one can regulate by addressing one substance or one company at a time. The relation between the actor and the reactor, i.e. nature, is in most cases, as

---

[16]Tengström, E., (1996) 'The Necessity and Difficulties of a Dialogue between Natural Scientists and Social Scientists', in Rolén (1996).

[17]Minna Gillberg, *From Green Image to Green Practice – Normative action and self-regulation*, Lund Studies in Sociology of Law 6, Lund University, 1999.

[18]Cf. Lars J. Lundgren and Jan Thelander, *Nedräkning pågår. Hur upptäcks miljöproblem? Vad händer sedan?*, (Count down is taking place. How are Environmental Problems detected? What is happening afterwards?) Naturvårdsverket informerar, 1989.

Staffan Westerlund has underlined, not a linear one.

To conclude, we have a knowledge deficit in the field of environment, within both natural and social sciences, which makes legal regulation problematic. It is hard to set up material rules telling what to do. The law instead falls back on competency rules pointing out who is entitled to take the decision, such as the organization of the environmental courts, and procedural rules telling us how to act when taking the decision, as for instance the regulation on environmental impact assessment. The environmental law is made for comprises. Another consequence is that the interpretation of the content of law becomes a cognitive instead of a normative issue. The application of law therefore to a large extent will be in the hands of other professions than lawyers.[19] Finally, the complex issues we are dealing with when we are talking about relations in nature and the concept of sustainable development, leads us to an increased attention to perceptions of risk and limitations in science.[20]

Law also has to be legitimate in order to fulfil its functions. If the law is not legitimate it is not followed spontaneously and the implementation will entail heavy costs for control. There are several factors which influence environmental legislation in a negative way from a legitimacy point of view. One of these factors is related to what has just been said about problems in relation to the knowledge deficit. Environmental problems are often invisible. One might notice the consequences but not the causes, which makes it hard to find the actor responsible. Since environmental problems are due to external effects and not that much an effect of individual wrongdoings, they cut across normal legal rationality and it is thereby hard to make legal claims. Environmental problems are something that follow as a side effect of something else you want to achieve. They are in a way systemic and follow from the way the economic system as it is structured.

Environmental legislation belongs to a category of norms which the American sociologist and game theoretician, James Coleman, has called dis-joint norms.[21] Coleman makes a distinction between what he calls con-joint and dis-joint norms.

---

[19]The influence of knowledge or knowledge based reality descriptions is something which in late modernity calls for more and more observance in relation to legal decision-making. See Inger Johanne Sand, 'A Future or a Demise for the Theory of the Sociology of Law: Law as a normative, social and communicative function of society', in Retfærd (Justice) vol. 90, 2000, p. 55. See also Christian Joerges et al. (eds), *Integrating Scientific Expertise into Regulatory Decision-Making*, 1997.

[20]See Richard Young, *Uncertainty and the Environment. Implications for Decision Making and Environmental Policy*, Edward Elgar Publishing, Cheltenham 2001. Young articulates the Shackle theory and evaluates it with a case study in relation to disturbance of an ecosystem.

[21]James Coleman, *Foundations of a Social Theory*, Harvard University Press, Cambridge. Mass. 1990.

The underlying idea is the difference between the addressee and the beneficiary of the law. Normally law consists of con-joint norms in the sense that addressee and beneficiary are identical. Coleman mentions, however, a ban on public smoking as an example of dis-joint norms. The ban itself has the smoking public as addressee. The beneficial group, though, is those who not smoke and who want to be protected against 'passive smoking'. Environmental regulation obviously belongs to the same category. It has its focus on the polluters to the benefit of those who are involuntarily exposed to the pollution. As a consequence, the legitimacy ascribed to the law by those who are expected to refrain from something, furthermore at some expense, may be expected to be little compared to what the beneficiaries of the law think about it. Environmental regulation therefore in general can be said to operate in an unfriendly milieu and cannot for that reason be expected to be adopted spontaneously. It has to be complemented by controlling authorities, such as the County Administrative Board and the Environmental Protection Agency. This is a common characteristic of all intervening norms, to which environmental law belongs.

Another problem with the efficiency of environmental law is related to the fact that environmental issues are politically unique. Politics has always been about distributive conflicts, class against class, state against state, sex against sex, ethnic groups or social groups against each other, etc. Environmental problems do not have this background. It is another matter that the environmental problems hit some – most often vulnerable – groups more than others. These consequences, anyhow, probably just add to the differences that occur between individuals and groups in society independent of environmental questions. This means that there are no natural bearers of the environmental interest in society. It affects us all as much or as little. This makes even the legitimacy of environmental law delicate. The ordinary way of dealing with political issues, creating alliances on a political level in order to form compromises, is not that easy.

Furthermore, the environment to a large extent is without ownership which makes it hard to fall back on self-interest in a regulatory strategy. Air and the sea are without private ownership. Garret Hardin has elaborated on this aspect in terms of tragedy of the commons, i.e. that no one has incentive to take precautionary measures to safeguard commonly shared property.[22] The same goes for the problem of pollution. No one has the incentive to refrain from polluting commonly shared non-property. Here one can talk about making use of economic theory, a kind of inverted free-rider situation. There seems to be no spontaneous willingness to safeguard nature as long as it does not create any benefits for the actor. The

---

[22]Garret Hardin, 'The Tragedy of the Commons', 162 *Science* 1243–48, and Garret Hardin and John Baden (eds), *Managing the Commons*, San Francisco, 1977.

initiative to create so-called bubbles may be viewed as a means to compensate for the lack of ownership.[23] The effect on nature, however, seems to be marginal in those few cases where it has been practised. After all, ownership in these cases is about the right to pollution, something which must, from the perspective of environmental protection and sustainable development, be regarded as a capitulation.

We thus have not only a knowledge deficit but a legitimation deficit as well. It is hard both to regulate in a substantive way and to create legitimacy for compliance with the law. There are some basic presumptions that lie behind the compromise and control paradigm dominating the legal strategies in the field of environment, presumptions which we have to change. To make a difference, we think that we have to start from a reverse angle in order to find strategies for a sustainable society, which is the same as identifying sustainable strategies. Here we face a need for a paradigm shift. Every paradigm according to Thomas Kuhn, is built on certain basic assumptions.[24] Interestingly enough, these are never openly expressed. They represent hidden values, still functioning as axioms for the paradigm. It is the same in the paradigm forming environmental legislation. The compromise and control paradigm builds on a reactive (command) instead of a proactive (demand) strategy.

In the compromise and control paradigm the industrialists are regarded as unreliable crooks who do not want to do anything voluntarily for the environment. This means that people are treated with distrust and thereby reproduce this mentality. For the same reason, and for reasons related to what has been said about dis-joint norms, environmental problems give rise to minimalistic reactions. You never do more than you have to and you even withhold information in order to 'escape'. To illustrate this statement we would like to quote a former CEO of a large Swedish corporation:

> The essence of Swedish industrial tradition was to put up as much resistance as we possibly could against all environmental restrictions. A strategy that the company had practised was to slow down the proceedings by consistently resisting all measures by claiming that they were technically impossible and not economically feasible, even though we knew the technology worked. Another efficient strategy was to provide the Licensing Board with insufficient and vague material and always apply for exemptions. And one could always appeal the decision to the Government which generally was rather lenient to the industry. It was a behaviour which was an integral part of Swedish business life. And I have to admit that in the beginning I did not reflect about the ethical

---

[23]Lena Gipperth, *Miljökvalitetsnormer. En rättsvetenskaplig studie i regelteknik för operationalisering av miljömål (Environmental quality norms. A legal science study in regulating techniques for operationalisation of environmental goals)*, Uppsala 1999.
[24]Thomas Kuhn, *Scientific Revolution*, (1962).

dimension of this resistance because it was the normal conduct, and a conduct which also was allowed by the law through the great potential to get exemptions.[25]

So far the established environmental regulation paradigm has legitimized the exploitation of the environment and, at best, managed to create some sort of limiting frames. In a sense, we may talk about a responsive law, but with a misdirected responsiveness towards an outdated instrumental rationale, that of the unsustainable industrial society. Governmental policy and legislation have failed to create incentives for industry to alter their practice by constantly undermining its own governmental green image through the substantial possibility to receive far too generous environmental licensing terms. This old paradigm builds on the stick instead of the carrot, on state regulation instead of self-regulation, on compromise and control instead of demand and development. This type of presustainable industrial instrumentality can never promote and push a sustainable development forward due to its ethical inflexibility and normative closure. Therefore the normative forces in society that work for the benefit of the environment cannot influence the reasoning and practice of such legislation and decision making. And these environmental normative forces must therefore find other ways of channelling ethical preferences to alter the practices of society.[26]

The Emergence of a New Paradigm

From a negative to a positive valuation of the environment[27]

The basic assumption underlying the rationale of the compromise and control paradigm is that the environment has a negative value, i.e. negative economic connotations which hamper growth. During the 1990s it became evident that a paradigm shift within industrial behaviour was taking place (see Wolff and Zaring, this volume). It was a shift in which the concept of the environment finally gained positive connotations for industry. This surprising shift in attitudes evolved very quickly. It is a rapid development, which stands in stark and total contrast to what the law and its present tools is capable of handling. Furthermore, it is a shift in industrial practice that our authorities could not anticipate, comprehend or actually

---

[25]Minna Gillberg, *From Green Image to Green Practice – Normative action and self-regulation*, p. 173, Lund Studies in Sociology of Law 6, Lund University, 1999.
[26]Minna Gillberg, *From Green Image to Green Practice – Normative action and self-regulation*, chapter 9, Lund Studies in Sociology of Law 6, Lund University, 1999.
[27]Ibid.

believe in. The implications are that this development has made most theories of governance and regulation strategies outdated. In other words, this event has to a certain degree rendered contemporary environmental legislation obsolete.

In order to grasp the explosive and positive development we can detect on the environmental arena, regarding eco-labelling and the implementation of environmental management systems, it is necessary to take a step back in time. 'Officially' everything started in 1962 when the biologist Rachel Carson published her epic book *Silent Spring*. It is a book that describes the negative impact that the increased and uncontrolled use of pesticides have on the natural environment. It should be noted that Rachel Carson's work had begun already in 1945 when she wrote to *Reader's Digest* and proposed an article about the effects of DDT on nature. The proposal was refused, but her early observations concerning DDT and pesticides in general laid the foundation for the work she embarked on at the end of the 1950s, resulting in the publication of *Silent Spring* in 1962.

The book had the effect of a gunshot which triggered the modern environmental debate and inspired the creation of environmental movements all over the USA, and the world. The environmental organizations were very much coloured by the spirit of the time, following in the tracks of the civil rights movement. These new organizations could be exemplified by the Public Interest Research Groups (PIRGs) which were created all over the USA, inspired by the activities of the consumer advocate and environmentalist Ralph Nader. The approach of the groups was that they should 'bring about an initiatory democracy' and create a 'full-time citizenry, aware, informed, and constantly acting upon social, economic, and political institutions in the public interest'. For these groups, and other similar movements which in one way or another were the result of the same social climate, the political agenda and the driving force said that things could indeed be changed. This intensive North American environmental debate spread to Europe, primarily through scientists who transcended both national and scientific borders.

In Sweden the early stages of a similar development can be attributed to such actors as Georg Borgström, Hannes Alvén, Hans Palmstierna and Björn Gillberg.[28] Through their dedicated books and active participation in the societal discourse, these 'whistle blowers' contributed to the foundation of the multi-faceted environmental movement that developed during the early 1970s.

The development was a process that focused on knowledge, awareness and social commitment and responsibility. It was within this context that the 'environmental' consumer for the first time became a visible agent of power. In 1971 a television programme showed how a young scientist, Björn Gillberg, washing an oily shirt

---

[28]George Borgström, *Gränser för vår tillvaro* (1964), Alfvén, H. and K. (1969), *M-70.*, Hans Palmstierna, *Besinning* (1972), and Björn Gillberg, *Mordet på framtiden* (1973).

spotlessly clean by using a coffee creamer. He did this to illustrate the danger with the chemicals we put in our food that end up in our bodies. The chemical substances which made up the coffee creamer had the same functions as those used in ordinary detergents. Due to the enormous impact of the media (only two television channels existed in Sweden at the time) the coffee creamer 'Prädd', which was owned by Astra, more or less lost its entire Swedish market within a couple of weeks.

It was the consciousness-raising environmental debate of the seventies, which focused primarily on energy consumption, food additives and pesticides, that resulted in the green parties of the eighties and also formed the knowledge base for the environmentally aware consumers of the eighties and nineties. Consumers who deliberately use their wallet as an efficient tool to influence both the production and the supply of goods and services utilize the market mechanisms in their refusal to buy certain goods. The growing environmental consumer movement of the nineties has clearly proved itself to be a most powerful agent of change.

What appears to make all the difference is that the environment has finally gained a positive financial value (marketing value) for the production sector thanks to these environmental consumers. Environment has become a competitive factor. This can partly be explained by the fact that consumers' demands have managed to shift the focus toward a more sustainable consumption. However, the effectiveness of this demand, as a means of pressure, is naturally heavily dependent upon the continuous awareness of consumers articulating their preferences. Environmental NGOs have played (and still play) a decisive role in the process as providers of information and knowledge, and in initiating these normative collective actions.

Another contributing factor of great importance, closely related to the work of the environmental NGOs, is the industrial sector's change in attitude towards environmental issues. This change, besides the consumer demands, can be attributed to the many environmental scandals and the lack of good will which has struck Swedish industry in recent decades. The increased environmental awareness in media reporting and the general appreciation (value) of environment in society have contributed to the ethical labelling of companies that misbehave as being environmental crooks.

The increased ethical pressure created by a well-educated public body has forced the industrial sector to come to terms with the fact that a step from 'green image' to 'green practice' is a necessity to secure their future position in the marketplace.[29] In addition, the communication that has taken place between NGOs and industry, for instance, before the Licensing Board, has forced the active environmental NGOs to accumulate extensive knowledge about technological solutions for environmental problems. Therefore it is not surprising that industry

---

[29]See Minna Gillberg, (diss.), *From Green Image to Green Practice. Normative Actions and Self-regulation*, Department of Sociology of Law, Lund 1999.

today turns to these environmental NGOs to ask for advice. From the conflicts in the courts or before the Licensing Board, industrial companies know that the NGOs master the environmentally sound 'technology' in arguing for the principles of best available technology (BAT) and best environmental practice (BEP). The great difference today is that we have managed to create a context where these old enemies can shift their traditional position, based on the procedure before the Licensing Board, to a position where they meet in order to co-operate towards a common goal, namely, to reduce the environmental impact of the company.

In this 'peace' process it has also become evident for the industrial sector that there is money to be made in environmentally friendly technology, since in general it is also a cost-effective measure. This is due to the inherent nature of these technical solutions, which often aim at a reduction of energy consumption, the use of raw materials and the polluting emissions.

It appears that we can attribute the present development of eco-labelling and environmental management systems to the operation of two forces acting in conjunction: the demands from environmentally aware consumers and, not least important, the industrial sectors' own internal needs to monitor their operations in order to avoid environmental scandals and the financial losses following in the wake of such accidents. A telling example is the case of Skanska, the company responsible for the catastrophe with the Hallandsåsen tunnel in south Sweden.[30] Within two weeks, from the day the matter was reported by the media, Skanska's stock market value fell by over SEK 2 billion. Financial analysts referred this strong market reaction to a distrust of the way Skanska had handled environmental matters. It is quite clear that the scandal could have been avoided if Skanska had asserted that an environmental management system had been implemented in a proper way. With a well-functioning management system it would have been impossible for the product that caused the scandal, Rhoca Gil, to slip through the demands put on product control and purchasing.

The magnitude of the environmental problems has now reached a level which is exerting pressure on the producers and other actors whose activities have negative implications for the environment. Environmental problems have now become so apparent that actors in the marketplace are forced to take ecological considerations into account in their decision making if they want to stay in business. Through these market mechanisms, our environment has gained a positive and crucial value for the companies when it comes to decision making in marketing, investments and other financial matters. The industrial sector now views the natural environment as

---

[30]See about this case, Håkan Hydén, 'The Dependency of Laws upon Norms – The Hallansdsås Debacle', in Helle Tegner Anker and Ellen Margrethe Basse (eds.), *Land Use and Nature Protection. Emerging Legal Aspects*, Copenhagen (2000), pp. 351–376.

a positive asset instead of a negative factor as it did previously. This paradigmatic shift has led to proactive behaviour in response to a demand, as opposed to reactive responses to legislation imposed by the state.

From Compromise and Control to Demand and Sustainable Development

As expressed above, the consumers' perception of environmental problems plays a vital role in the determination of the link between individual purchases and their environmental impact. It was the consumers with this perception who created a considerable increase in the demand for environmentally friendly products in Europe at the end of the 1980s, chiefly in Germany and Scandinavia. This demand created a situation where a growing number of market actors wanted to purchase products and services that were documented as environmentally friendly. At the time environment-related product information was not available to any great extent. However, the market responded to this new demand with its customary speed, by starting to mark products as 'environmentally friendly'.

In a real explosion, a great variety of labels sprang up with declarations about the environmental quality of the product, for example, that they were 'biodegradable', 'natural' or 'ecological'. The labels often signalled environmental friendliness by portraying everything ranging from seals and trees to fields in bloom. A majority of these labels were 'self-declarative' and were not initially controlled by any external party. Several national eco-labelling schemes were also introduced during this time. The oldest national scheme is the German Blauer Engel which dates from 1979. Today the Engel offers over 4,000 products and is the largest eco-label scheme in the world. At present approximately 20 national and private schemes exist in the world, but there is no ongoing global harmonization process between these different labels (in contrast to the environmental management systems).

Parallel to this development outside the political and legal system, there has been a shift in focus in the political rhetoric during the 1990s from the environment towards sustainable development. Due to space constraints we presuppose here that the background to this is known.[31] This shift has clear normative consequences.[32]

---

[31]For a discussion about research strategies in support of sustainable development, see Forskningsrådsnämnden, rapport 1998:17, *Möjligheter och hinder på väg mot faktor 10 i Sverige* (*Possibilities and hindrances for factor 10 in Sweden*). See also L. Anders Sandberg and Sverker Sörlin (eds), *Sustainability, The Challenge. People, Power and the Environment*, London 1998.
[32]This has been elaborated in an article, 'Hållbar utveckling ur ett normvetenskapligt/ rättssociologiskt perspektiv', (Sustainable Development from a Perspective of Norm Science/ Sociology of Law), in Olof Wärneryd and Tuija Hilding Rydevik (eds.), *Hållbart samhälle – en antologi* (*Sustainable Society – an Anthology*), Forskningsrådsnämnden Rapport 1998:14, pp. 138–164.

The concept of environment is in a way value-neutral. It does not imply anything. It can include bad or good consequences for the environment. Most scientific and practical work in relation to nature has a negative impact on nature, but we usually then do not call it environmental science. When applying this label to scientific work we seem to have a common understanding of something in favour of the environment. When we talk about sustainable development, the concept in itself implies certain value standpoints in an environmentally friendly direction. Even the understanding of the concept of environment is affected.

Environmental problems may be comprehended as dealing with what nature can stand. In this view the natural laws put restrictions on the societal laws and what man can do without destroying the conditions for the reproduction of nature. The concept of sustainable development also implies that the societal systems should be able to be reproduced. Not only the ecological systems, i.e. the physical and biotic systems, but also the economic, political and socio-cultural systems should be designed in a way that makes reproduction possible. In this perspective nature and society are integrated in the concept of environment. The common denominator when we are looking at the environment from a sustainable development perspective is that it includes all living conditions. The concept of environment consists of the living conditions that surround us, ecological, economic, socio-cultural, etc. A good environment is, in other words, equivalent to good living conditions. The concept of sustainable development, thus, brings us from a reactive to a proactive approach to the environment.

The introduction of the perspective of sustainable development thus has revolutionary consequences if it is taken seriously. We have seen that the perspective has been given a prominent place in the Swedish Environmental Code which starts with the following statement: 'The purpose of this Code is to promote sustainable development which will assure a healthy and sound environment for present and future generations.' We have also seen that this advanced objective is followed by contradictory statements later on in the Code, and this gives way for an application of the law which is not that promising. The question is whether there are any possibilities of strengthening sustainability in a legal perspective.

The concept of sustainability strengthened legal factors and interests promoted and protected by law in relation to other competing factors, such as economic and political interests. It is not enough to say that all factors should be taken into account in a balancing decision-making process when deciding what is sustainable and what is not. There must be some weighing principles. In this process of interpretation some general principles function as guidelines. Natural laws are immutable, in contrast to societal laws which are an outcome of compromises. Natural facts are what the philosopher John Searle calls brutal or objective facts, in contrast to societal facts which are institutional or subjective facts.[33] Garret Hardin makes a similar distinction between first and second order of truth.[34] With this view the concept of sustainable development puts the emphasis on the nature in the balancing process with societal factors. Taking rights seriously[35] would, for instance, in the wording of the Swedish Environmental Code, result in a priority for ecological values in any conflict with other interests. At least it means that the principle *in dubio pro natura* must be valid, i.e. that nature has the preference of interpretation.

In order to operationalize one can make use of the Pareto optimality parallel. An act is then sustainably optimal if it meets the following criteria. As long as someone's life improves from an act in accordance with the social, economic and political system without impinging on the ecological systems in a negative way, that act is sustainably optimal. It is another thing that the principle might be hard to apply due to lack of or at least unreliable knowledge. The same objection could, however, be made to the Pareto optimality principle.

The problem lies in how to implement this mentality of sustainability. When it comes to legal strategies researchers and practitionars have to change paradigm, find other means of approaching the environmental problems. The interest of future generations cannot easily be advocated in a legal process. It is not realistic to expect that such a gigantic problem as how we treat nature can be solved by legal or other means of force. If the normative systems in society do not favour the environment, there are small chances of success in any regulating endeavour. It is too costly in many respects. Nor is it, for the same reason, realistic to calculate with economic incentives in the form of green taxes or fees in order to create sustainability.[36] Of course one can gain certain advantages.[37] These means of

[33] John Searle, *The Construction of Social Reality*, 1995.

[34] Garret Hardin and John Baden (eds), *Managing the Commons*, San Francisco, 1977.

[35] Cf. Ronald Dworkin, *Taking Rights Seriously*, Cambridge, Mass. Harvard University Press, 1978.

[36] Thus Runar Brännlund and Bengt Kriström in their article in Thomas Sterner (ed.) *The Market and the Environment. The Effectiveness of Market-Based Policy Instruments for Environmental Reform*, Edward Elgar Publishing, Cheltenham 1999, assert 'that there appears to be little room', based on the experience of the recent Swedish Green Tax Commission. See also Runar Brännlund and Ing-Marie Green (eds), *Green Taxes. Economic Theory and Empirical Evidence from Scandinavia*, Edward Elgar Publishing, Cheltenham 1999. Cf. Mikael Skou Andersen and Rolf-Ulrich Sprenger (eds), *Market-based Instruments for Environmental Management. Politics and Institutions*, Edward Elgar Publishing, Cheltenham 2000.

[37] See Thomas Sterner (ed.) *The Market and the Environment. The Effectiveness of Market-Based Policy Instruments for Environmental Reform*, Edward Elgar Publishing, Cheltenham 1999.

influence can be used to prevent degradation but not to create sustainability. The most important factor for a sustainable society is to change the driving forces and the motives for individuals and companies acting in relation to nature.

To sum up, the point of departure for a new paradigm is the changing entrance value from negative to positive when it comes to the approach to environmental issues. With the environment becoming a competitive factor, it acquires positive connotations. It becomes something desirable instead of something which should be criticized. In this way the foundation is laid for an interplay between self-regulation and state regulation.[38] Environmental aspects change to be a self-interest for the companies to take into account in the decision-making processes. The value of sustainability becomes an integral part of the overall strategy of many enterprises. We can here talk about sustainable enterprises.[39]

The development which we have outlined above will of course also have consequences for the work of the governmental authorities and for the possible use of law as an instrument of environmental change. Even if legislation as such has never been a prime mover in the field of the environment, it has had an important role as a levelling factor. In the name of competitive neutrality, the law has helped to block one company from trying to gain comparative advantages by not caring about environmental effects. This role of environmental law had a dynamic effect in combination with the self-regulating activities we have mentioned. The most important parts of the present legislation are perhaps about environmental impact statements. On the whole, the procedural rules are the most highly developed in the Swedish legislation. These administrative routines are of direct relevance for making environmental projects aware of environmental risks. Another way of strengthening the role (rule) of law would be to open up for public participation in the decision-making processes to a larger

---

[38]Rena I. Steinzor asserts that '(t)he most prominent alternative to command and control is industry self-regulation', see Rena I. Steinzor, 'Reinventing Environmental Regulation: The Dangerous Journey From Command to Self-Control', in *Harvard Environmental Law Review* 1998, p. 104.

[39]Cf. Robert Isaak, *Green Logic. Entrepreneurship, Theory and Ethics*, Kumarian Press, West Hartford, Conn., 1999. In August 1991, McKinsey & Company published an ambitious poll of the attitudes of international business leaders toward the environment. According to this, business commitment to global environmental improvement is widespread and deep-seated. For example, 92 per cent of respondents agreed with the statement that '(t)he environmental challenge is one of the central issues of the 21st century', and 82 per cent agreed that 'there is a need to assume responsibility for one's product even after it has left the plant'. 67 per cent agreed that 'pollution prevention pays'. Rena I. Steinzor asserts that '(t)he most prominent alternative to command and control is industry self-regulation'. See Rena I. Steinzor, 'Reinventing Environmental Regulation: The Dangerous Journey From Command to Self-Control', in *Harvard Environmental Law Review* 1998, p. 156.

extent.[40] Furthermore, promoting individual rights in environmental law could be an important upgrading of the environmental interest. In addition to this, the possibility to make claims for damages should be developed. Obviously the lesson from environmental law history is that claims for damages can work in a preventive way. This might require a shift in legal culture where the courts to a larger extent are willing to put more emphasis on individuals than on authorities and industry. There seems to be signs in this direction.[41] These proposals for alterations in environmental law are in accordance with the paradigmatic shifts we have identified and elaborated upon above.

Today, the innovative place of action, though, is the shadow of law, not the law in itself. What we are witnessing seems to be the dawn of a new paradigm in policy making where the role and importance of the state, on a national/local level, is fading with regard to the actual implementation of environmental objectives. Therefore, a future strategy for the active state might be to try to locate, support and take advantage of the strong normative structures and movements in society, which could work for the benefit of the environment.[42] It is a matter of finding a strategy for an environmentally efficient use of the limited time and the limited governmental resources available for the purpose of sustainability. The market as a self-referential organism is naturally not to be trusted to an infinite extent, but, where it can act as an enforcer, we must take advantage of it and exploit its survival mechanisms of self-regulation.[43]

---

[40]The situation seems to be similar in UK. See William Wilson, *Making Environmental Laws Work. Law and Policy in the UK and USA*, Oxford 1999. Wilson considers the political context in which environmental laws are made, and the implications for long-term public support of them. He examines the ways in which the law-making processes in Britain and in Europe effectively exclude public participation and offers suggestions for ways to change these processes with examples of American alternatives. The US seems to have a more developed system of public participation. See Rena I. Steinzor, 'Reinventing Environmental Regulation: The Dangerous Journey From Command to Self-Control', in *Harvard Environmental Law Review* 1998, p. 172. Steinzor argue for instance in the article that 'EPA must address the question of funding the participation of public interest representatives at a level that is more that a manipulative parody of what it would take to ensure equivalent bargaining power', (p. 198), something which must be regarded as an advanced system.

[41]A promising example is the judgement from the Environmental Court in Växjö, where the court for the first time in Swedish environmental legal history took a standpoint contrary to the recommendation of the Government by prohibit mining (Case No 399-99). The Government had taken a decision according to the Mineral Act.

[42]For an example of this reasoning relating to education, see Per Wickenberg, (diss.) *Normstödjande strukturer. Miljötematiken börjar slå rot i skolan*, Lund Studies in Sociology of law 6, Lund 1999.

[43]Minna Gillberg, (diss.), *From Green Image to Green Practice. Normative Actions and Self-regulation*, Department of Sociology of Law, Lund 1999.

In the last part of this chapter we will now try to bring together the lessons learned over time with the ambition of trying to create a sustainable mix of regulatory strategies.[44] The crucial question, as regards the choice of regulation strategies, can be formulated as follows: how and in what circumstances, and in what combinations, can the main classes of policy instruments and actors be used to achieve optimal policy mixes? Certain combinations of instruments are either inherently counterproductive or, at the very least, sub-optimal. Furthermore, the outcome will depend on the particular context in which the instruments are combined, not to mention the underlying rationale of decision making in the political and legal system (an issue we also will address in the following section).

A Sustainable Regulation Strategy of Governance: the Law of Sustainable Development – Integration in Practice through Sustainable Management Systems (SMS)[45]

Strategies for sustainable governance – towards a law of sustainable development

Today it is clear that traditional legal methods have exhausted their capabilities and cannot master, or regulate, the complex (environmental) problems of our times. We could also say that our legal culture – with its roots in roman law, supplemented by the continental theory of sovereignty after the Treaty of Westphalia (1648), the German theory of legal personality of the State and the Anglo-Saxon tradition on the rights of the individual – has come to its limits when trying to address and regulate the common future of humanity and earth. Legal science has for paradigmatic reasons never been interested in the outcome of law, or the evaluation/assessment of itself. The social processes whereby law has been produced and their effect on people and nature have been of no interest to conventional legal theory, or to

---

[44]We thereby follow the line that has been staked out by Neil Gunningham and Peter Grabosky in their important book, *Smart regulation. Designing Environmental Policy*, Clarendon Press, Oxford 1998. The authors have scanned practice and potential in environmental regulation worldwide, pointing out strengths and weaknesses with a variety of regulatory instruments, such as command and control regulation, self-regulation, voluntarism, education and information instruments, economic instruments such as property rights, market creation, fiscal instruments and charge systems, financial instruments, liability, performance bonds, deposit refund systems, removing perverse incentives and free market environmentalism.

[45]This chapter is based on Minna Gillberg's research regarding the concept of Governance and Sustainable Regulation Strategies and her work with the development of a Sustainable Management System based on ISO 14000. Professor Ludwig Krämer, head of Unit, DG XI, provided invaluable and important information regarding the topic and the development within the EU.

lawyers. Those 'irrelevant and subjective' matters have been left for political science or sociology. The main task of law and jurisprudence has been limited to the 'objective' interpretation of rules, irrespective of the political source/ideology of the rule. As discussed in earlier chapters, the underlying rationale of our environmental legislation has been determined by an economic rationale where the political considerations have been to ensure growth and the protection of industry against legal claims and prohibitions. Environmental legislation has not primarily been installed to protect the environment, but to ensure a regulated and licensed exploitation of the environment. Hence, it is not surprising that traditional environmental law is incapable of promoting the realization of sustainable development. Nor is it surprising that the modern state has lost its governing role in relationship to market forces and the protection of the environment.

As we have seen so far, the state has not been very active in trying to locate, support and take advantage of the strong normative structures and movements in society which work for the benefit of the environment. Nor has the state paid much attention to the possibilities of exploiting the market and its survival mechanisms of self-regulation. So, the question then arises if the national state, the EU and other global institutions of human good should continue on the path where the subject of sustainable development is downplayed, a path where people wash their hands and blame the internal market, the USA or the WTO for their own inactivity. This is where one have to stop for a moment and return to Agenda 21, a document that was adopted and signed by every country in the world. Its legal value lies in the sense that it expresses worldwide consent for the strategy of sustainable development and thus incorporates precepts, general principles and practices of international law and compels the governments of national states to enact legal rules of sustainable development.

For the first time states and the global community are faced with the task of having to reshape, in accordance with Agenda 21, our society in order to ensure its/ our survival. We stand far away from the point where the liberal state (or any other state) can afford to reject governmental 'maternalism/paternalism'. We can no longer afford to reject the ideal of the classical Aristotelian coercive state, which strives for the achievement of a virtuous democracy, a just polity that ensures justice and survival. Sustainable development *per se* requires that the state reassumes its governing role. The rise of normative consumer movements calls for the state and the global community to assume responsibility for a just reformation of society and the global community order. Sustainable development is not just a change of the rationale behind economic policy, but in fact a planned social reform to be attempted under the aegis of the state governed by law. A brief analysis of Agenda 21 shows that the objectives of the document go far beyond 'mere economic policy and in fact aim to reform society so as to make it more just. So that, in our times, survival is impossible without justice, whereby a justice of sustainable development is not merely the social policy exercised until now'.[46] To succeed with the implementation of Agenda 21, the state has to commit itself to certain values which it renders legitimate and obligatory for all, and on the basis of those values the state must plan and implement substantive changes of society and its institutions.[47]

Agenda 21 leaves no doubt that its directives recommend 'the building up of the governing capacity of the State in every country, especially in developing countries. In that way, the effectiveness of the State is seen as its ability to govern. Consequently, instead of the hitherto neutral State which follows the lead of the market, Agenda 21 calls for a sustainable and capable State which:

a) will establish law enforcement mechanisms and periodically check their efficacy.

b) upholds control over the market, and recognizes that the market is accountable and responsible for its actions, and that the market is not and cannot become an *equal* partner with the state in the enterprise of sustainable development.[48]

However, the legal statutes of sustainable development do not exist; they have to be created. We do not yet know what the appropriate institutions will look like, or the appropriate behaviour of citizens. We also know that Agenda 21 provides us with general guidelines and principles which must create the basis for the shaping of future statutes and behaviour. Hence, from a legal perspective, change has to be brought about through a broader perspective of the role that law should play in promoting sustainable development. It is necessary for jurisprudence to go through a process of self-criticism, i.e. to acknowledge and analyse the fact of failure, in order to create a new identity that can meet the challenges of a complex and threatened world.

What we are implying here is a shift of perspective and rationale for the law, i.e. a break with the 'right to do' and the licensed right to exploit nature. Instead we should use a concept coined by Michael Decleris, namely, 'the Law of Sustainable Development'. What Decleris proposes is a certain fundamental starting hypothesis on the sustainable relationship between humanity and nature (man-made systems and ecosystems), which can be described by a finite number of general principles deriving from Agenda 21. However, these principles of sustainable development cannot be procedural but must be substantive and value-based, because only then can they lead in a particular direction, i.e. that of sustainable development.[49]

[46]Michael Decleris: 'The Coming Systematised State of the 21st Century', in Tercera Escuela de Sistemas, Valencia 1995.

[47]Ibid.

[48]Ibid 47.

[49]*The Law of Sustainable Development*, a report produced for the European Commission by Michael Decleris, J.S.D. (Yale), Justice, founder and President of the Fifth Section of the Greek Council of State (the Greek environmental Supreme Court), 2000.

An important question we need to address is that all public policies must be interrelated and harmonized around the prime factor of concern for the conservation of natural, cultural and social capital.[50] How, then, can we by legal techniques create a by-pass that can override the short-term and growth-oriented economy and the whims of unsustainable politics, and at the same time act as an integrating mechanism for all policy and decision making? It appears as though the only solution lies in a general requirement for a constitutional clause of sustainable development, in conjunction with a material right for the public to invoke such a clause in a constitutional court. A constitutional clause of sustainable development would force the legislator, courts and authorities to 'undertake that responsibility and fulfil it, abiding by the prime duty of conferred upon them directly by the Constitution to protect the environment and the quality of life'.[51] This approach would also ensure the development of a context-dependent and problem-oriented jurisprudence. In effect, the legal implications for, say, industrial activities, would most likely be that no activities that could be deemed unsustainable should be allowed, i.e. only activities that were rendered as promoting sustainable development should receive a license. The burden of proof regarding sustainability would naturally be assigned to the party applying for a permit or license.

The establishment of a new Law of Sustainable Development requires the adoption of a supra-legislative system of general principles and criteria for sustainable development. These principles, based on Agenda 21 and other MEAs, would act as a guiding check-list, linking the constitutional clause with the legislator, detailed regulation and legal practice. In a highly interesting report to the European Commission, Michael Decleris draws up twelve principles of sustainable development which could serve as a constitutional basis for policy making and legislation.[52]

The incorporation of such principles in national legislation and policy making would substantially strengthen the establishment and effective promotion of sustainable development, since the underlying rationale and the driving force of the law would be to promote activities that are sustainable (and not to promote unsustainable activities, as the present legal system does). In addition, these principles, particularly the twelfth principle, requires that the system of law must be transformed into an open and flexible system in continual communication with societal development, for example, normative movements and the development of new progressive technology. The four pillars that must form the basis for a new legal system based on *sustainable democracy*, are the principles of transparency,

[50]Ibid 50.
[51]Ibid.
[52]Ibid.

access to information, public participation and the principle of accountability (applied not only to corporations and citizens but also to the state, authorities and the legal system itself).

An active state must also become better at creating smart regulation strategies and supporting structures in order to promote an industrial behaviour that is proactive in response to a demand (as opposed to reactive in response to traditional top-down legislation imposed by the government). Such a state must actively identify and support an industrial practice that is advanced in relation to sustainable development.[53] Furthermore, since present political decision making, particularly in the field of environment, relies on facts, which are determined by natural science, a governing and sustainable state can no longer leave natural science to play all alone – especially since natural science, like legal science, has been hiding behind its scientific demands of 'objectivity'. The concept of sustainable development no longer allows for objectivity in an epistemic sense, meaning that natural science can no longer remain 'objective' in relationship to the development of society and its research focus. In other words, science and research have to take stance for sustainable development. Along these lines, sustainable development requires integration not only of society, but also of the research community. We cannot create a sustainable society without a research approach that integrates and develops criteria and indicators for all three dimensions of sustainable development. However, this integration of research does not happen automatically, and the attempts so far have for various reasons failed. To deal effectively with the complexity of environmental problems it might be a necessity to establish 'task force working groups' (*ad hoc* groups), composed in such a way that they comprehensively cover all aspects of a particular environmental problem (including research and practice). The main objective for these groups would be to analyse and suggest practice-oriented measures for different threats to the environment, for instance, the threats listed by the Swedish National Environmental Protection Agency. By integrating theory, technology, and practice, this interdisciplinary model could lay the foundations of a lasting 'green' strategy for the management of environmental problems.

It is probably necessary to create a forum around each environmental threat. A forum that on one hand gathers scientific and medical competence and data for the identification of the character of the problem as an environmental and health problem, and on the other hand can also provide social scientific competence (including behavioural science and the humanities) for an analysis of the causes of

---

[53]For examples of advanced industrial practice, see for instance Minna Gillberg, (diss.), *From Green Image to Green Practice. Normative Actions and Self-regulation*, Department of Sociology of Law, Lund 1999.

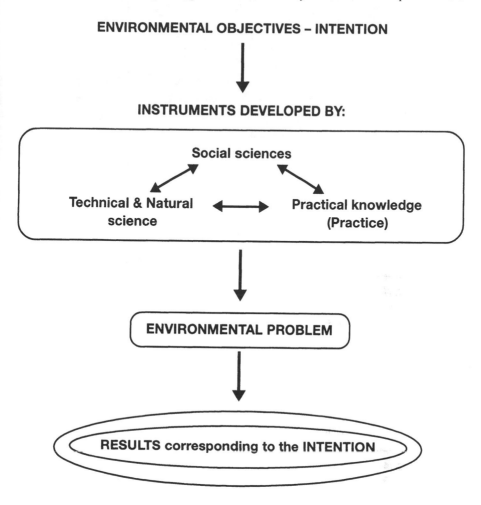

Figure 1   The task force approach

the environmental problem and for the identification of the problem in the phase when the current instrument of control is applied. In addition, we need technical knowledge to be able to decide on appropriate technical solutions. Finally, but not least important, there is the need to have persons in the *ad hoc* group who have practical expert knowledge in the field within which the control is operating. These persons should monitor the entire process, and make sure that the different parts of the process of control are coherent and that they work in practice. It is not completely uncommon that proposed technical solutions appear ideal from the point of view of the prerequisites they are based on, but when they face reality, additional unforeseen problems occur that thus reduce the effect of the control measure, or delay it. In conclusion, the integration of theory and practice, suggested by the model, could perhaps serve as a proactive basis for future research strategies regarding the management of environmental problems. The environmental Task force groups, could develop practice-oriented action programmes with a focus on concrete measures, i.e. the objective would be to suggest sustainable alternatives to human practices which hampers sustainable development.

Integration in Practice through Sustainable Management Systems

One of the major problems with the concept of sustainable development is, as we have discussed throughout the article, the demand for integration and the interrelationship between the three dimensions of sustainability. The previous sections within this chapter have in the main focused on how to alter rationality for the practice of society's systemic and governing features; we shall now turn to the sphere that lies outside the realm of our governing structures. In the context of sustainable development and a global market which knows no borders, it is of the utmost importance to alter the practice of financial organizations, such as international corporations.

As discussed in section 5.2, recent development indicates that an important paradigm shift has taken place within industrial practice as regards environmental issues. This paradigmatic shift has paved the way for a proactive behaviour in response to the concepts of sustainable development and Corporate Social Responsibility (CSR). The issue of CSR has for several years received a considerable and growing attention, both in international academic literature and among important actors within the business community.[54] Mr Björn Stigson, president of the World Business Council for Sustainable Development (WBCSD), describes this shift with the following words: 'In business there is a strong shift from the situation only some years ago, when the environment and sustainable development were seen singularly as problems and risk factors by Business, to a situation where they today clearly are seen as opportunities – sources of efficiency improvement, technological development and competitive advantage.'[55]

As a matter of fact, many corporations and business organizations have gone

---

[54]J.B McGuire, A. Sundgren and T. Schneeweis, 'Corporate social responsibility and firm financial performance', *Academy of Management Journal*, 31: 854-872, and B. M. Ruf, 'The development of a systemic, aggregate measure of corporate social performance', *Journal of Management*, Jan-Feb, 1998.

[55]Speech held by Mr Björn Stigson, president of WBCSD, at the EU symposium on The EU Sustainable Development Startegy, 23rd of Febuary 2001, Stockholm.

further than considering these issues. Today the global business community is striving towards the development of tools, criteria, indicators and standards for sustainable management and development. Just to mention a few examples there we have the Dow Jones Sustainability Index launched in 1998 which foremost serves as a benchmark for sustainability-driven investments accounting for a total market capitalization of approximately USD 5 trillion (or, 19.1 per cent of the Dow Jones Global World Index). There is also the Global Reporting Initiative (GRI) which was established in 1997 by a partnership between an NGO, the Coalition for Environmentally Responsible Economies (CERES), and United Nations Environment Programme (UNEP). The mission was to develop and disseminate applicable 'Sustainability Reporting Guidelines' for voluntary use by organizations reporting on the economic, environmental, and social dimensions of their activities, products and services. In June, 2000, GRI released the revised version of the Sustainability Reporting Guidelines of 1999 which had been piloted by some 20 multi-national companies from around the world. Furthermore, in January 1999, UN Secretary-General Mr. Kofi Annan called on world business leaders to 'embrace and enact' a set of nine principles (UN Global Compact), promoting sustainable development and human rights (formally launched in September 2000), in their practice and to support complementary public policy initiatives. In addition, the Organization for Economic Co-operation and Development (OECD) released the OECD Guidelines for Multinational Enterprises which provide voluntary principles and standards for responsible and sustainable business conduct as regards; disclosure, employment and industrial relations, environment, combating bribery, consumer interest and the concept of science and technology (S & T). In so far, two management standards for the social dimension of sustainable development have been developed, to assist a structured implementation of these issues in the daily management of an organization.[56] However, these management systems have not been altogether successful and are not in widespread use, mainly due to the fact that they do not build upon the existing and predominant environmental management systems ISO 14000 and EMAS. The response from the business community has been a reluctance to implement a parallel management system which operates with different routines from the environmental management system. Therefore, only a couple of hundred companies have been certified according to these standards in comparison to the thousands that are certified according to ISO 14000 and EMAS.

In several respects we may conclude that the business community has come further in trying to define the meaning of sustainable development than the national

---

[56]The two standards, Social Accountability 8000 (SA 8000) and AccountAbility (AA 1000), were respectively developed by the Council on Economic Priorities Accreditation Agency and by the Institute of Social and Ethical Accountability.

state, or for that matter, the EU. It is a situation which is symbolically promising, but, in the light of governance, there are several questions to be raised. Governments and the EU have left the field open to industry, to elaborate and define sustainability and corporate social responsibility (CSR) on their own terms and conditions. The exploitation of the market's self-regulating mechanisms is, as we have seen, essential, and this approach has been extremely successful in a Scandinavian context. However, when it comes to the creation of a Sustainable European Union, politics cannot continue to leave the initiative to the (single) market forces. It is time for policy makers to initiate co-operation with the progressive part of industry, in order to leave the laggard industrial dinosaurs behind (an example of this is the UN initiative Global Compact). We need to find strategies that can break the lobbying of traditional industry and we need to find arguments that can neutralize the trade lawyers who run the show in Brussels. The answer lies in identifying those corporations which have understood that, in order to play the game of the future, they need to change their strategies, and make a clean break with the old industrial paradigm of resistance, in which the environment had negative connotations. This is where the importance of creating supporting structures for sustainable development comes into the picture. A smart government would seize the opportunity of the day, and develop those efficient instruments that already are accepted by industry, and for the better, in widespread practice. Here, we are referring to the environmental management systems ISO 14001 and EMAS.

We have developed a Sustainable Management System by integrating criteria for the social dimension into the existing standard of ISO 14000.[57] The criteria for the social dimension builds upon the following international documents; Agenda 21, Global Compact, ILO Conventions, ILO Tripartite Declaration of Principles concerning Multinational Enterprises and Social Policy and OECD Guidelines for Multinational Enterprises. The result is a consistent, concrete and comprehensive management system. An organization which implements such a Sustainable Management System is forced, not only to evaluate its impact in relationship to nature, but also in response to its social impact. The great advantage of our integrated management system is that the systemic approach clearly explicates the immediate and 'symbiotic' links between the social, ecological and economic dimension of an organization, and put the focus on the inter-relationship with the surrounding world, i.e. the sustainable (or unsustainable) impact on society and nature.

Since an environmental management system defines legislation as the minimum criteria for the organization's environmental performance; the task is to set and

---

[57]Minna Gillberg who has been carrying out practice oriented research on this matter in co-operation with SKANSKA (a world-wide building and construction group) and KPA (a major pension fund in Sweden).

achieve goals that lie beyond the legislators demands. (The overall performance and the set goals of a certified organization are periodically assessed by an accreditation organization, and if the goals are not met the organization unconditionally will lose its certificate.) This means that we can conclude that companies that are EMAS registered, or ISO 14001 certified, have an environmental performance that supersedes present legal demands. In other words, these companies invest in the environment and a sustainable future on a voluntary basis, placing more stringent demands on themselves than any existing ad binding legislation.

As a telling example, several European and US based companies that are ISO 14000 certified have committed themselves to implement the Kyoto-protocol (and beyond) and furthermore, abandon fossil fuels in favour of bio-based fuels. Companies like Shell and BP has committed in reducing greenhouse gas emissions by 10 per cent below 1990 levels by 2002 and 2010, respectively.[58] Car companies like Ford, DaimlerChrysler, Toyota and Honda have committed to contribute to the decarbonization process by mass producing fuel cell cars (for bio-based methanol) and placing them on the market starting 2002/2003. It is estimated that the CO_2 emmisions deriving from cars will be reduced by at least 30 per cent within ten years due to the introduction of fuel cell driven cars and bio-based fuels. This important step in the decarbonization process does not stop at cars but goes beyond, to the highly significant sector of energy production. The year 2001 Ballard Power Systems delivered their first 250-kilowatt commercial and stationary fuel cell unit. To bluntly conclude what we argue here, there is no law or multilateral agreement, which would be compatible with the rules of the EC internal market and other international trade laws, that could force industry to abandon of fossil fuels. Only consumer demands on sustainability and ethics can bring such a change about, and this development is what we are witnessing today.

In this perspective, environmental law has played a minor, if not irrelevant, role. Herein lies our persistence (and interest, as researchers in sociology of law) to dwell upon this new phenomenon, an evolution of voluntary standards establishing new norms, codes of sustainable industrial conduct, which as it seems, have a stronger impact on industrial behaviour than the traditional way of codifying norms through legislation. And, where the enforcement mechanism is not constituted by penal law and other forms of punishments but of various forms of gratitudes and rewards; an alternative regulation strategy which builds upon the carrot (demand) instead of the stick (command), and to our surprise, it actually works.

In conclusion, environmental management systems have introduced a new paradigm of proactive and self-regulating industrial behaviour as regards environmental issues, particularly, concerning resource efficiency. Hence the

[58]S. Dunn, *Decarbonizing the Energy Economy*, State of the World 2001.

wording in section 5, that the place of innovative action is found in the shadow of the law. In consequence, it would not be unlikely if the creation of a Sustainable Management System, under the auspices of the European Commission and/or the International Standardization Organization (ISO), would have similar societal implications as regards the promotion of sustainable development, i.e. that it would prove to be an instrument that could bring sustainable development into industrial practice. And, most importantly, it would be a tool of governance which can act as a bridge between policy-making (legislation) and the progressive global business community, i.e. a concrete tool for sustainable governance which builds upon transparency and co-operation – a proactive interplay where 'law acts as a safety belt and the market, as a cost-efficient enforcer' which can further and promote sustainable development.[59] It is, as our introductory Chinese proverb indicates, crucial to develop feasible strategies for a change of course that are appealing to the multinational actors who (whether we like it, or not) determine our path, or, we will most likely end up where we are headed.

[59]M. Gillberg, 1999.

References

Alfvén, H. and K., (1969), *M-70*, Bonnier, Stockholm.

Andersen, M. S., and Sprenger, R-U. (eds) (2000), *Market-based Instruments for Environmental Management. Politics and Institutions*, Edward Elgar Publishing, Cheltenham.

Borgström, G., (1964), *Gränser för vår tillvaro*, Tema, LTs förlag, Stockholm.

Brännlund, R., and Green, I. M., (eds) (1999), *Green Taxes. Economic Theory and Empirical Evidence from Scandinavia*, Edward Elgar Publishing, Cheltenham.

Brännlund, R., and Kriström, B., (1999), 'Energy and environmental taxation in Sweden: Some experience from the Swedish green tax commission', in T. Sterner (ed.), *The Market and the Environment*, Edward Elgar Publishing, Cheltenham.

Christensen, J., (2000), *Rätt och Kretslopp. Studier om förutsättningar för rättslig kontroll av naturresursflöden, tillämpade på fosfor (Law and Ecocycles. Studies of the preconditions for legal control of the flow of natural resources, applied to phosphorous)*, Uppsala.

Coleman, J., (1990), *Foundations of a Social Theory*, Harvard University Press, Cambridge. Mass.

Dunn, S., (2001), 'Decarbonizing the Energy Economy', in L. R. Brown et al. (eds) *State of the World 2001*, The World Watch Institute.

Dworkin, R., (1978) *Taking Rights Seriously*, Cambridge, Mass. Harvard University Press.

Gillberg, B., (1973), *Mordet på framtiden*, Wahlström & Widstrand, Stockholm.

Gillberg, B., and Templin, A., (1988) *Mord med statligt tillstånd (A governmental License to Kill)*, Wahlström & Widstrand.

Gillberg, M., (Diss.) (1999), *From Green Image to Green Practice – Normative action and self-regulation*, Lund Studies in Sociology of Law 6, Lund University.

Gillberg, M., and Hydén, H., (1998), 'The Swedish Environmental Legislation. Consensus in Favour of the Environment?', in H. Weidner (ed.), *Alternative Dispute Resolution in Environmental Conflicts*, WZB, Berlin: Edition Sigma, pp. 71–93.

Gipperth, L., (1999) *Miljökvalitetsnormer. En rättsvetenskaplig studie i regelteknik för operationalisering av miljömål (Environmental quality norms. En legal science study in regulating techniques for operationalisation of environmental goals)*, Uppsala.

Gunningham, N., and Grabosky, P., with Sinclair, D., (1998) *Smart Regulation. Designing Environmental Policy*, Clarendon Press, Oxford.

Hardin, G., (1968), 'The Tragedy of the Commons', *Science* 162: 1243–48.

Hardin, G., and Baden, J., (eds) (1977), *Managing the Commons*, San Francisco.

Hydén, H., (diss.) (1978), *Rättens samhälleliga funktioner, (The societal functions of law)*, Lund.

Hydén, H., (1991), 'Towards a Postinterventional Environmental Law', in *Jahreschrift für Rechtspolitik* 1991, pp. 245–259.

Hydén, H., (ed.) (1998), *Rättssociologiska perspektiv på hållbar utveckling, (Sociology of Law Perspectives on Sustainable Development)*, Sociology of Law Research Reports, 1998:1.

Hydén, H., (2000), 'The Dependency of Laws upon Norms – The Hallansdsås Debacle', in H. T. Anker and E. M. Basse (eds), *Land Use and Nature Protection. Emerging Legal Aspects*, Copenhagen.

Hydén, H., Wickenberg, P., and Gillberg, M., (2000), *Miljöledning i Citytunnelprojektet (Environmental Management in the Citytunnel project)*, Research Report in Sociology of Law, 2000:2.

Isaak, R., (1999) *Green Logic, Ecopreneurship, Theory and Ethics*, Kumarian Press, West Hartford, Connecticut.

Joerges, C., et al. (eds), (1997), *Integrating Scientific Expertise into Regulatory Decision-Making*, Nomos-Verl.-Ges., Baden-Baden.

Kagan, R., (1994), 'Regulatory Enforcement', in D. Rosenbloom and R. Schwartz (eds), *Handbook of Regulation and Administrative Law*, New York.

Kuhn, T., (1970), *The Structure of Scientific Revolution*, Chicago.

Lundgren, L. J., and Thelander, J., (1989), *Nedräkning pågår. Hur upptäcks miljöproblem? Vad händer sedan? (Count down is taking place. How are Environmental Problems detected? What is happening afterwards?)*, Naturvårdsverket informerar.

McGuire, J. B., Sundgren, A., and Schneeweis, T., (1988) 'Corporate social responsibility and firm financial performance', *Academy of Management Journal*, 31: 854–872.

Michael, D., (1995) 'The Coming Systematised State of the 21st Century', in *Tercera Escuela de Sistemas*, Valencia.

Palmstierna, H., (1972), *Besinning*, Rabén & Sjögren, Stockholm.

Ruf, B. M., (1998), 'The development of a systemic, aggregate measure of corporate social performance', *Journal of Management*, Jan–Feb.

Sand, I. J., (2000), 'A Future or a Demise for the Theory of the Sociology of Law: Law as a normative, social and communicative function of society', in *Retfærd (Justice)*, vol. 90.

Sandberg, A. L., and Sörlin, S., (eds) (1998), *Sustainability, The Challenge. People, Power and the Environment*, London .

Serres, M., (1990), *Le Contrat Social*, Paris.

Sörlin, S., (1991) *Naturkontraktet. Om naturumgängets idéhistoria*. Stockholm.

State of the World 2001 (2001), World Watch Institute, W.W. Norton, London.

Steinzor, R. I., (1998) 'Reinventing Environmental Regulation: The Dangerous Journey From Command to Self-Control', in *Harvard Environmental Law Review* 1998.

Sterner, Thomas, (ed.) (1999), *The Market and the Environment. The Effectiveness of Market-Based Policy Instruments for Environmental Reform*, Edward Elgar Publishing, Cheltenham.

Tengström, E., (1994) 'The Necessity and Difficulties of a Dialogue between Natural Scientists and Social Scientists', in M. Rolén (ed.) *Environmental Change. A Challenge for Social Sience and the Humanities*, FRN Report 94:3.

Utting, P., (2000), *Business Responsibility for Sustainable Development*, UNRISD, Occasional paper 2000:2, Geneva.

Wärneryd, O., and Hilding Rydevik, T., (eds) (1998), *Hållbart samhälle – en antologi (Sustainable Society – an Anthology)*, Forskningsrådsnämnden, Rapport 1998:14.

Westerlund, S., (1997), *En hållbar rättsordning, (A Sustainable Legal Order),* Göteborg.

Westerlund, S., *Aktuell miljörättsteori (Actual Environmental Law Theory)*, www.imir.com.

Wickenberg, P., (diss.) (1999), *Normstödjande strukturer. Miljötematiken börjar slå rot i skolan.* Lund Studies in Sociology of law 6, Lund.

Wilson, W., (1999) *Making Environmental Laws Work. Law and Policy in the UK and USA,* Oxford.

Young, R., (2001) *Uncertainty and the Environment. Implications for Decision Making and Environmental Policy*, Edward Elgar Publishing, Cheltenham.

Redirecting Infrasystems Towards Sustainability

Arne Kaijser

Introduction

One of the most fundamental societal changes in the Western World in the past two centuries has been the introduction and expansion of a number of large technological systems for transportation, communication, energy and water supply, and sewage and garbage collection. A common characteristic of these *infrastructural systems*, or *infrasystems* for short, is that they facilitate movements of different kinds; of people, goods and information.[1] Furthermore, they provide services that are publicly accessible and which fulfil a basic function in society.

On a macro level, infrasystems have brought about a transition from a basically 'nature-based' economy, where the location of industries and other economic activities was primarily dependent on the access to waterways and natural resources, to a 'culture-based' economy, where easy access to man-made infrasystems is decisive for the location of most economic activities. Infrasystems were first built between and within cities, and they have contributed to a fast urbanization. The expansion of infrasystems has furthermore enabled an intensified exploitation of natural resources as well as a division of labour on a hitherto unknown scale. As a result many production systems that previously were of a local or regional scope have become global in scope. In short, the development of infrasystems is an important factor behind the sustained economic growth of the past two centuries. For everyday life, infrasystems have implied what could be called a 'convenience revolution'. Many of the most strenuous household tasks have been taken over by electric household appliances, tap water, and central heating. Furthermore, the car and the telephone have given many households a dramatic increase of mobility and reach. Until now, it is primarily in the industrialized world that infrasystems have had these effects. In the developing countries most infrasystems are only accessible

[1]For a discussion of these concepts, see Kaijser 1984 and Kaijser 1999.

to the relatively wealthy, but the poor are eager to attain them as well. An expansion of infrasystems in these countries will most probably have tremendous impacts in coming decades.[2]

The success of infrasystems can be summarized in the words: cheap, convenient and reliable. However, it is this combination of advantages that is also the root of their environmental problems. Through the ease and cheapness of their services, infrasystems have strong tendencies to encourage increasing consumption of scarce resources. It is literally like opening a tap of water; why bother about the amount of water you use in the shower when it is so easy, so pleasant and so cheap? Infrasystems have affected the environment in two ways: First, many of them have severe *direct* consequences for the environment. Just think of the emissions from motorcars, airplanes and power plants. Secondly, infrasystems have considerable *indirect* consequences for the environment. The increased capacity for mobility they have brought about has enabled many households to settle in relatively large dwellings in suburbs, and they have developed increasingly energy- and transport-intensive lifestyles.[3]

Most of the articles in this volume deal with mental, legal and other immaterial frameworks forming barriers for people changing their habits and lifestyles in an environmentally sustainable direction. This article focuses instead on material frameworks, which also have a strong influence on the behaviour of people, and which constitute important and long-lasting obstacles for changes towards sustainability. Furthermore, these material frameworks have a high degree of inertia over time. The infrasystems and the built environment in which we live today are the result of decisions and efforts made decades and even centuries ago. Likewise, decisions and efforts we make to build and rebuild systems and structures will shape the material world for future generations.

It is the strong historic legacy of infrasystems that is the point of departure of this article. I believe that a prerequisite for redirecting infrasystems towards sustainability is an understanding of their developments in the past and of their influence on settlement patterns. The purpose of the article is to contribute to such an understanding. The article covers a very broad topic and is therefore of a rather general nature.[4] I try to highlight some patterns and mechanisms that I believe are particularly important, and for pedagogic reasons I use a number of examples as

---

[2]This article deals with the Western World and focuses therefore on the *re*direction of infrasystems. However, I believe that the developing world could learn much from the historical experience in the Western World when building and expanding their infrasystems.
[3]Jonsson et al., 2000.
[4]As the article covers such a broad area, I have been restrictive when it comes to references. It is partly based on previous works by me alone and in co-operation with others, and in these works more extensive references can be found.

illustrations. The structure of the article is as follows. First a short introduction to the research field is given and some characteristics of infrasystems are presented. The following section focuses on the dynamics of infrasystems, analysing the factors and mechanisms that have contributed to their development in the past. The third section briefly outlines the impacts of infrasystems on society at large and especially on settlement patterns. The last section is devoted to a discussion of what lessons we can learn from history about the possibilities for redirecting infrasystems in the future.

Infrasystems as a Field of Research

The concept 'infrasystem' is used to denote a certain category of large technical systems, and the study of infrasystems partly falls within a research field that is called the Large Technical Systems approach (sometimes abbreviated LTS). A major impetus for the development of this field was a book published in 1983 by the American historian of technology, Thomas P. Hughes, entitled *Networks of Power*, which analyses the establishment and growth of electricity systems in the United States, Germany and Britain. Hughes regards electrical systems in a broad sense, as *socio-technical systems*, including not only technical components, but also the people and organizations that design, build and operate these components, as well as the legal and economic frameworks for these activities. Fundamental in his approach is to analyse how electricity systems in this broad meaning attain their specific form in the environment in which they develop. He tries to clarify the dynamics of these systems, and which kind of actors and problems have been essential in different phases of their development. By comparing electricity systems in three countries he shows that they acquired varying *styles* in different countries and regions, due to the specific geographical, political, economic and cultural conditions of each region.[5]

Hughes' book inspired many other scholars, and since the mid 1980s large technical systems have become a field of research, attracting a growing number of historians and social scientists. Many of them have also made comparative studies of a certain system in several regions or countries.[6] The concept 'large technical system' is very general, and my point with introducing the concept infrasystem is to focus on a certain kind of system, namely those which fulfil a basic function in

[5]Hughes 1983.
[6]See for example the anthologies Mayntz and Hughes 1988; La Porte 1991 Braun and Joerges 1994 Summerton 1994; Blomkvist and Kaijser 1998; Coutard 1999; Andersson-Skog and Krantz, 1999.

society by facilitating movements of people, goods and information, and which are publicly accessible. I believe that a fruitful way to study infrasystems is by making comparisons among different kinds of systems. This has not been done very much within the LTS-tradition. Such comparisons can give us a better understanding both of the specific properties of individual systems and of the common characteristics of all infrasystems. And by combining comparisons between countries and between systems we can get insights into how infrasystems are shaped by nation-specific political and cultural traditions on the one hand, and by system-specific technical and economic properties on the other.[7]

Characteristics of Infrasystems

A common characteristic of infrasystems is that they facilitate movements of different kinds. They can be described as consisting partly of a *network* of links (like rails) and nodes (stations), partly of a *flow* passing through this network (trains). *Distributive systems*, like electricity, water and television, have a unidirectional flow from one or several central nodes to a large number of users. *Accumulative systems*, like sewer and garbage collection, have a reverse unidirectional flow, from many users to one or several central nodes. *Communicative systems*, like telecom, post and transport systems, provide a two-way flow.[8]

The character of the networks (in particular the links) varies greatly between different systems. Some systems presuppose the construction of *specific networks* consisting of, for example, electric lines, water pipes and rails, which are built solely for the particular system. (These systems are sometimes referred to as grid-based systems.) Other systems are largely based on *natural 'networks'* like water, air or electro-magnetic waves in conjunction with harbours, airports, transmitters and receivers. And still others, like the post system or the internet, use *existing transport or communication networks* in combination with terminals, post-boxes, servers and the like. Another categorization of networks can be made in terms of their geographical shape, and in particular the points of access for the users. *Point-shaped networks* are accessible to the users only in a limited number of exclusive nodes, i.e. airports, stations and harbours. *Line-shaped networks* are accessible along their links where nodes can easily be arranged, i.e. telephone and electricity lines. *Surface-shaped networks* i.e. radio, TV and mobile phones, are accessible in every point within a distribution area.[9]

One fundamental aspect of an infrasystem is its *reliability*. As they fulfil basic

[7]Kaijser 1999.
[8]Jonsson 2000.
[9]Kaijser 1994.

functions in society, which are necessary for many different kinds of activities, interruptions or breakdowns can have far reaching consequences. It goes without saying that it is utterly important to prevent major accidents leading to loss of lives and property, as such accidents can undermine the societal credibility of a system. But it is also essential to minimize small interruptions or delays. If these become frequent, users lose confidence and may change to other, competing systems. The American sociologist Charles Perrow makes a useful distinction between *tightly coupled* and *loosely coupled* systems. A tightly coupled system is more vulnerable as a disturbance in one component rapidly spreads to other parts, while a loosely coupled system has more redundancy. To obtain a sufficient degree of reliability in a tightly coupled system, it is often necessary to have one or several system operators co-ordinating the flows. In a loosely coupled system, like road traffic, the establishment – and enforcement – of common rules and norms can often result in an acceptable reliability.[10]

Another important aspect of infrasystems, not least in relation to the issue of sustainability, is what economists call their *external effects*, which can be both positive and negative. If a large airport is built outside a city, for example, this will normally give considerable positive economic effects for the whole city region, attracting new businesses etc. Often new office buildings surround the motorways leading to a major airport. A large part of these economic effects do not result in profits for the airport or airline companies, and are therefore called 'external' effects. Such desired effects have often motivated public authorities to contribute to the investment in airports and similar facilities. But an airport will also have considerable negative effects, not least in the form of noise disturbance for people living in its vicinity. These effects are also external in the sense that they often do not cause any costs for the airport company. Thus the establishment and expansion of infrasystems will have different kinds of effects for different actor categories. And these effects are not distributed at random; they reflect political and economical inequalities in a society. In general, the wealthy and politically influential citizens get more of the positive benefits, while the poor get more of the negative ones.

The Dynamics of Infrasystems

In the introduction I argued that the success of infrasystems can be summarized in the words: cheap, convenient and reliable. This is, to be sure, an observation in hindsight. When efforts have been made in the past to develop and establish new infrasystems, there has always been a major uncertainty whether the system would

[10]Perrow 1984.

be viable or not. Most historical research has focused on the successful attempts, but it should be stressed that uncounted attempts have actually failed. In this section I will first outline some characteristics of three phases in the development of individual infrasystems, and then briefly discuss the interplay between infrasystems.

Establishment

It is well known that many infrasystems are based on a radical technical innovation, connected with inventors such as Alexander Bell, Thomas Edison, Guglielmo Marconi and many others. However, the establishment of the system on a first market generally requires a huge investment, and at this early stage it is mostly very difficult to assess the future demand for the services of the system. The establishment phase is therefore characterized by a very high uncertainty. A crucial problem is how to find mechanisms for overcoming this uncertainty. A common feature of many systems is that a fundamental *institutional innovation* was made in an early stage, enabling a common use of the new system by many different groups, thereby diminishing the uncertainty. (I use the concept institutional innovation in a rather wide sense, referring to a change in the relation between provider and user of a service often accompanied by a change in the nature of the service.) I will briefly outline the introduction of gas lighting to illustrate this process.

In the late eighteenth century a number of engineers and inventors tried to develop new technologies for lighting. At the turn of the century a Frenchman, Phillipe Lebon, and an Englishman, William Murdoch, independently designed simple gasworks. Their machines produced gas out of coal, peat and wood, which were much cheaper than the dominant light sources at this time, tallow and whale oil. However, a gasworks represented a considerable investment in retorts and pipes. It is therefore not surprising that the first commercial use of gas lighting was in factories. In 1802, Murdoch installed gas lighting in the premises of the Boulton & Watt Company in Birmingham, and in the following ten years gasworks were installed in about a dozen other English factories.[11]

For the owners of large factories, needing to light huge buildings, gas lighting implied a significant reduction of their lighting costs. However, for other categories of light consumers, the high capital costs of a gasworks represented an insuperable obstacle. Thus the market for the new lighting technology seemed to be restricted to large factories with a high demand for lighting. It was at this point that a radical new idea was developed by the German entrepreneur Friedrich Albert Winzer. His idea was to sell gas, not gasworks.

[11]This example is based on Kaijser 1986 and Elton 1958.

Winzer had built a small gasworks, imitating Lebon's plant, and had travelled across the continent trying to sell it, with little success. He came to London in 1803 and there he gradually came to the conviction that a precondition for gas lighting to reach a big market was that the investment cost for a gasworks could be distributed among many users, thereby reducing the 'entry fee' for each of them. He developed a plan to establish a joint-stock company that would build a big gas-producing plant in the middle of London, and a whole network of pipelines under the streets. This would enable the company to sell gas at a relatively low price to a large number of subscribers (even though the cost to install pipes within a building still posed an obstacle) and also to supply gas for street lighting. For many years Winzer tried to convince capitalists, politicians and the public of the reliability and profitability of this plan. His ideas met fierce opposition not least from people with interests in the supply of existing means for lighting, whale oil and tallow. Finally he succeeded to form an alliance of actors that was sufficiently strong, and in 1812 Parliament gave permission to found the 'The Gas Light and Coke Company'. Two years later the company began selling gas and within ten years gas lighting was used by thousand of subscribers in shops, restaurants, workshops, offices and households as well as for street lighting. Many other cities in Britain and on the continent followed London's example in the next decades.

The idea of selling gas instead of gasworks led to an institutional innovation of fundamental importance. It was by finding a way for a communal use of the expensive gas producing plants that gas lighting became affordable for many more. In short, gasworks became an infrasystem, and the introduction of gas lighting led to a radical change of urban life in the course of the nineteenth century. A similar story – of a communal use of a system by many different kinds of groups – can be told for a number of infrasystems. In the railway system a key innovation was to provide not just a rail (like the canals did) but to offer transport, both of passengers and of goods.[12] For achieving a fast expansion of the postal and telegraph systems it was essential that many kinds of interests (civil service, armed forces, railway companies, businessmen, newspapers etc.) could share the same service. In the case of urban water systems a prerequisite for mustering the necessary capital was that the water could be used for several purposes; in households, in factories and for fire fighting.

---

[12]The first public railway, built in 1825 between Stockton and Darlington in England, was organised like a canal, just offering a rail. This proved to be inefficient when traffic grew, and the second public railway built in 1830, between Manchester and Liverpool was therefore organised in a totally different way; it offered transport, not just a rail. In addition to building a railway, this company also bought wagons and locomotives and organized train traffic following elaborate timetables. See Lilley 1973 and Kaijser 1994.

My point is thus, that the crucial problem in the establishment phase of an infrasystem is uncertainty, and that the establishment of a new system has generally involved an institutional innovation, which has enabled communal use of heavy investments making a new service affordable for many different categories of people. Moreover, this innovation has to be supported by an alliance of powerful actors. When one city or region has been able to establish a successful infrasystem, many others will soon want to follow its example. However, each city, region or country will try to adapt the institutional set-up of the infrasystem to its own political and socio-economical conditions.

The institutional shaping of an infrasystem can be seen as the result of an encounter between technology and society. The technical subsystem imposes certain demands on, for example, the degree of coordination and control, but these demands can be met within a more or less broad spectrum of organizational and legal frameworks. Which of these possible frameworks that is imposed depends on social and cultural traditions and the relative power of different interest groups in a country or a city. The institutional frameworks shaped for the first infrasystems in a city or a country have often served as a model when infrasystems have been established later on. This 'institutional transfer' has led to the emergence of specific national institutional regimes for infrasystems. In some countries public authorities have taken a very active role in the building and operating of infrasystems, while in other countries public authorities have primarily had a regulatory role, trying to prevent system operators from abusing a monopoly situation.[13]

Expansion

Once an infrasystem has been established and reached a first major market an entirely new situation develops. The revenues from sales provide an economic base and the experiences of building and operating the system often lead to further technical improvements. This shapes the prerequisites for the expansion of infrasystems, either by way of new customers wanting to use its services (outer expansion) or through an increase of the consumption by old customers (inner expansion). There are generally strong economic and social forces for expansion.

Let us first look at the economic forces. The marginal costs for providing additional units of service have usually decreased in expanding systems, due to *economies of scale* and *economies of scope*. The economies of scale arose primarily on the 'production side'. For example, the production cost for a unit of gas or electricity decreased when the size of the generating plants increased.[14] Likewise,

---

[13]See Kaijser 1999 and Dobbin 1994.
[14]Hirsh 1989.

the cost per passenger or unit of goods usually decreased as the size of ships, trains and airplanes increased. Falling costs enabled lower prices, which raised demand and spurred further increase of scale etc. The economies of scope arose primarily on the consumption side. For example, gas was first used mainly for gas lighting, which implied that most of the gas was used in the morning and in the evening. Huge gasometers were needed to store gas produced during the day. In the late nineteenth century gas was also introduced for cooking, water heating and for industrial processes and motors. These markets had a different demand pattern over time, and as a result a more even use of gas was achieved. Another way to phrase this is to say that the *load factor* of the system increased. A high load factor implied a better utilization of the huge capital costs of the system, and thereby lower costs per unit of service. Similar increases of the load factor have taken place in many other systems as well, when new categories of customers or new types of services have been introduced. This has often been achieved through very deliberate policies from the system operators offering the new users favourable prices.[15] Thus, while economies of scale arose through an increase of the size of the technical components of a system, economies of scope were achieved through a better system architecture, a more efficient balance among the components within the system.

In infrasystems providing communicative services there is also another kind of economic force stimulating expansion that I would like to call *economies of reach*.[16] The economies of reach have to do with the extent of the network, and thereby the number of people or places that can be reached by using it. It implies that the growth of a system can be an important quality in itself. In the telephone system, for example, the connection of a new subscriber was not only rewarding to the new subscriber, but also to all the existing subscribers, who received an additional person to call. This phenomenon was sometimes deliberately exploited. At the turn of the century there were two competing telephone companies in Stockholm, and they tried to increase the attractiveness of their networks by offering telephone subscriptions for free to doctors, pharmacists and other professionals that their subscribers wanted to reach.[17] Also when local telephone systems were interconnected by interurban lines, this implied a dramatic increase in the number of people that could be reached, which in turn spurred additional people to become subscribers. Economies of reach have also been of importance

[15]Hughes 1983.
[16]The term that is usually used to denote this phenomena is 'network externalities', but I think that the expression 'economies of reach' gives a better intuitive understanding of the phenomena.
[17]Heimbürger 1931, Helgesson 1999.

for the growth of transport systems. For example, the attractiveness of having a car increased as the network of roads grew and was improved. In many countries, powerful motorcar lobby organizations arose, which succeeded to persuade authorities to invest in the improvements of roads.[18] This paved the way for a fast increase in car ownership, which in turn gave an economic base (through taxes on cars and on fuel) for further investments in the expansion of roads.

Besides the economic forces there have also been strong 'social' forces for expansion. On a macro level expanding infrasystems acquired what Thomas P. Hughes calls *momentum*.[19] Companies, which had invested much capital and other resources in a system (both equipment producers and owners and operators of systems), developed a strong interest in the further expansion of the system. Furthermore, in many systems there was a need for engineers with special skills, and these often had a common educational background. Such professional groups often formed tightly knit networks sharing the same values and with similar views concerning the desired future direction of the system. They thus developed a common *system culture,* to use a concept introduced by Hughes. At many times a close co-operation among many influential actors towards a common goal shaped a strong force for expansion.[20]

On the micro level there have been parallel social processes. When a new infrasystem was established there was often a certain resistance towards it from potential customers using other means to fulfil the same function. Besides the costs for joining the new system, their resistance was often based in a lack of knowledge about how to use the new system or a reluctance to change habits and routines, for example changing from cooking on a wood stove to cooking on a gas stove. However, if their resistance was overcome, and they did change to the new system, they developed new skills and new habits making them dedicated followers of this system and reluctant to change to other ones, for example electric stoves.[21]

Thus, once an infrasystem was established on a first major market, strong forces for expansion arose, producing a spiral of growth. Economies of scale, scope and reach led to falling costs and decreasing prices of the services. This spurred the recruitment of new customers and the increase of consumption among the old ones, which led to a further decrease of costs etc. However, a fast expansion of a system was seldom a smooth process. Technical obstacles and difficulties often appeared, threatening to block the expansion. For example, increases of scale or distance have often been difficult to achieve. Thomas P. Hughes uses a military metaphor to

---

[18]Blomkvist 2001.
[19]Hughes 1983 and Hughes 1987.
[20]See for example Fridlund 1999.
[21]Hagberg 1986.

describe this phenomenon. He talks about a 'reverse salient' in an advancing front, which was a typical feature in the trench warfare during W.W.I. When such reverse salients emerged in the expansion of infrasystems, no resources were spared to try to overcome them. The best scientists and engineers available were engaged in these efforts, and radically new components, sub-systems or system-architectures were often the result, which in turn enabled a continued fast growth.[22]

A fast expansion of an infrasystem, brought about by alliances of powerful actors, often resulted in considerable economic gains for many parties. However, in many cases the system also led to negative effects in the form of pollution and health problems affecting other parties. For example, the coal-fired gas-producing plants gave many workplaces and households access to a convenient energy source, but it also brought about severe health problems, in particular for the workers employed in the plants and for people living close by. These plants also contributed to severe environmental degradation, and still today, dangerous chemicals often heavily contaminate locations of previous gas plants. Another example is the introduction of water closets in the apartments of well-to-do urban households at the turn of the century. This improved the hygienic conditions for the users of the closets, but it also contributed to the pollution of rivers and lakes, which in turn affected many other people. Thus, while expanding infrasystems were often extremely effective in solving problems for some people, they usually also contributed to the creation of other problems affecting other, less influential people. In the past such environmental and health problems have generally not become reverse salients in Hughes' sense, as they have not been acute threats to the further expansion of the systems. And therefore they have not attracted the same kind of attention as the direct obstacles to growth.

Stagnation

Expansion processes do not go on forever. Sooner or later infrasystems reach a phase when growth rates diminish and a phase of stagnation commences. One factor contributing to stagnation has been a weakening of economic forces for expansion. Economies of scale have reached a saturation level in a number of systems. The maximum size of power plants, freight ships and aeroplanes, for example, has hardly increased since the 1970s. Likewise economies of scope are usually more or less exhausted once the load factor of a system reaches a certain level. Diminishing economies of scale and scope have sometimes been combined with a saturation of demand for the services of infrasystems. For example, in many industrialized countries a rather fundamental change 'from bulk to bytes' occurred

[22]Hughes 1992. See also Fridlund 1999.

in the last quarter of the twentieth century. A growing part of economic activity was directed from material-intensive towards more knowledge-intensive products and systems, and this was reflected in a much slower growth of demand for services from energy, water- and transport systems. In addition, more efficient end-use technologies have also contributed to a slower growth of demand.

When it comes to economies of reach, it is not uncommon that systems have even experienced a transformation into *dis*economies of reach. In the case of road traffic in urban regions, additional cars have steadily increased the congestion on existing roads, and due to a scarcity of land it has been difficult to build additional roads. The more cars, the longer it takes for each to reach its destiny, and this of course hampers the further expansion of car traffic. A similar process affects aviation systems in some densely populated regions like Europe. In fact, high-speed trains are taking over parts of the passenger traffic between major urban centres in Europe, partly due to the frequent delays of flights.

There is one more factor, which has often played a crucial role in processes of stagnation and decline of infrasystems, namely competition from other systems providing similar services. (I will use the concept *functional equivalents* to denote two or more systems fulfilling the same function.) However, this is only one aspect of the interplay of infrasystems, and therefore it will be treated in the following section.

Interplay of Systems

There have been two main kinds of interplay among infrasystems. First competition among systems being functional equivalents has been a major stimulus for technical and economical improvements of systems, but also a cause for decline. Competition between gas and electricity systems is a clear example. Gas systems were uncontested during most of the nineteenth century as providers of a high-quality energy source, which could be used for lighting, cooking, heating and mechanical power. However, the electric power systems established in the early 1880s provided an energy source, which could be used for exactly the same purposes. In fact, when Thomas Edison designed his first electricity system for lighting in the Wall Street district in Manhattan he had the existing gas systems as a model. As a consequence a fierce competition arose between the promoters of the two systems, first for the lighting market and later on for the motor market, the stove market and the heating market, stimulating technical improvements of both systems. For example, the struggle for the lighting market led to a dramatic increase in the efficiency of both gas lamps and electric lamps, but in the 1910s electric lighting had become superior and gas lighting declined. Particularly in countries with abundant hydropower resources (and thus cheap electricity), gas systems were pushed back also in the other markets and many of them were closed down in the mid twentieth century. However, with the introduction of natural gas, the position of the gas industry versus electricity has been strengthened anew.[23]

Similar processes of competition have taken place among transportation and communication systems. In the second half of the nineteenth century canals competed with railways and half a century later there was an intense struggle between railways and motorcars. Likewise the telegraph and the postal systems had to struggle with telephone systems in the beginning of the twentieth century. The older systems struggled very hard to improve their efficiency, and while being pushed back in some markets they were sometimes able to keep their position in some market segment, in which they had special competitive advantages. Sometimes such a system has even been able to make a comeback. High-speed trains and electric tramways are two examples of this.[24]

There is also another form of interplay among infrasystems. They often play a complementary role to each other achieving synergistic effects. One classical example is the building of telegraph lines along railway tracks. The telegraph made it possible to communicate between stations and this made it possible to increase train traffic considerably. At the same time the railway facilitated the building and maintenance of telegraph lines, and also provided a guaranteed market for telegraph traffic. Another example of system synergism is a co-generation plant, in which the heat losses from electricity generation are used for the heating of many houses via a so-called district heating system. The combined production of electricity and heat is much more efficient than a separate production of each, and this was often a major incentive to build district-heating systems.[25] A third example is from the transportation sector. Transportation systems do not only compete. They often also need to co-operate because their networks have different coverage. However, an obstacle to such co-operation has been the high costs for trans-shipment. In the 1950s and 60s the container was introduced to facilitate the integration of different transportation systems, and thus to achieve synergistic effects.[26]

[23]Kaijser 1993.
[24]Grübler 1990.
[25]Summerton 1992; Hård and Olsson 1995.
[26]Egyedi 1996.

Infrasystems and Settlement Patterns

In the previous section, I have discussed the internal dynamics of infrasystems. In this section I will focus on the external effects of infrasystems on settlement patterns. I will argue that in certain periods the development of one or several new infrasystems have provided radically new opportunities for commerce, industry and housing. I call these periods *logistical revolutions* to emphasize the broad scope of changes that are brought about.[27] These periods of time can be seen as formative phases, when different combinations of infrasystems are possible to achieve. The power elites of some cities or regions will grasp the new opportunities earlier than others, and by establishing a fruitful mix of systems they will get a competitive advantage over other areas. Later on, their choices often become a model for other cities and regions. I will briefly outline some examples in space and time of such logistical revolutions.

During the sixteenth and seventeenth centuries a logistical revolution took place in Europe. In the previous centuries maritime trade had largely been confined in two regional trading networks: one around the Mediterranean Sea, and one in northern Europe around the Baltic and the North Sea. The development of better ships and the introduction of the compass and of accurate maps made it less dangerous for ships to travel past the Iberian Peninsula, and to cross the oceans to America, Africa and Asia. As a result, a truly global commercial system gradually evolved in the sixteenth century, with its first centre in Lisbon. The centre position was soon transferred to Antwerp, and a few decades later to Amsterdam. For more than a hundred years Amsterdam was the uncontested commercial and cultural centre of the world.[28]

Why did Amsterdam become the leading commercial centre in the late sixteenth century? It was in part due to very favourable geographical conditions. More important, however, was the ability of the leading citizens of Amsterdam to grasp the potential of the new technological and commercial opportunities. In the second half of the sixteenth century, the leading merchant families of Amsterdam initiated heavy investment in harbour facilities, and the city harbour soon had room for thousands of sailing ships. Furthermore, they stimulated the development of a shipbuilding industry in Zaandam, just outside the city, powered by large windmills. It specialized in building the 'fluitship', the best merchant ship of the time, and the Dutch merchant fleet became by far the biggest in the world. The ships did not merely bring goods to Amsterdam; they also brought merchants and sailors with information about distant market conditions. In 1600 or thereabouts, the city's

---

[27]See Andersson and Strömqvist 1988.
[28]Braudel 1992.

rulers founded the Amsterdam Bourse and a special Exchange Bank, and these institutions became the heart of the commercial activities in the city. All these infrastructural and commercial investments provided the basis for Amsterdam to become the centre of a trading network stretching over all the oceans of the globe.[29]

In addition to this long distance network, Amsterdam also became the centre of a local and regional transport system, which had no equivalent anywhere in the world. It was based on a very dense and fine-meshed network of canals and natural waterways, covering most of the Dutch Republic. These canals, which had been dug in the previous centuries for drainage purposes, facilitated local trade and thus encouraged a dynamic interaction between the towns and their agricultural hinterland. This spurred a remarkable urbanization process, and in 1600 half the population lived in towns and cities.

Furthermore, all these canals stimulated close interaction among the Dutch towns, which led to a specialization. Some of them focused on textile industries, others on breweries, dairies, shipbuilding etc. To facilitate travelling between these towns and cities, the Amsterdam rulers initiated a new public transport system in the seventeenth century interconnecting all major Dutch towns. Special canals were dug solely for passenger traffic by horse-pulled barges, or 'trekschuiten', which would make the traffic independent of the weather. Travelling by trekschuit was comfortable and cheap. They went at about 7 kilometres per hour and were very punctual, keeping to timetables. About one million passengers a year travelled on these barges in the mid seventeenth century, and Amsterdam was the central node in this network.[30]

In short, the ruling merchants of Amsterdam became the avant-garde of a logistical revolution based on a perfection of water-based infrasystems across the oceans and within their region, which made their city the commercial as well as the cultural centre of the world.

From Water-Based to Metal-Based Infrasystems

The Dutch had an exceptional transportation system *within* their country, due to special geographic conditions. But in most other countries inland transportation was slow and expensive, and formed a huge obstacle to economic development. In the mid nineteenth century, the building of railways and telegraph lines, opened up vast areas that had previously been virtually excluded from trade and commerce, and new towns and cities evolved along these networks of iron and copper. These new systems gave rise to a logistical revolution during which human built systems

[29]Ibid; de Vries and van der Woude 1995.
[30]de Vries 1981.

based on rails and lines of metal gradually replaced the previously dominant water-based infrasystems. Particularly in the United States with its vast distances and lack of roads, railways and telegraphs were of enormous importance, and Chicago became the railway city par excellence. Let me briefly explain why Chicago became so important.

The first efficient transportation route into the central parts of the United States was with steamboats along the Mississippi River and its tributaries in the 1820s. St. Louis, situated at the confluence of the Mississippi and the Missouri, became the most important trading town. When Chicago was established in 1830 it was also chosen from the water transport point of view. With the new Erie Canal it was possible for steamboats to go from New York to Chicago via the Great Lakes, and in the rainy season it was possible to travel with a canoe from Chicago along a small river across the watershed down to the Mississippi. In 1848, the leading merchants of Chicago founded the Chicago Board of Trade, which became a very important organization for the development of the city. When railways were being built in the mid nineteenth century westward from New York, Philadelphia, Boston and Baltimore, the Chicago Board of Trade was able to convince a number of railroad companies that Chicago would be the ideal end station to the west. At the same time, the Board of Trade organized the building of a whole number of small railway lines and canals to the west of Chicago which spread out, fanlike, over the prairie.[31]

The Chicago merchants started to organize a large-scale exploitation of the natural riches of the prairies to the west and of the forests to the north. Grain, animals and lumber were brought to the city in ever-larger quantities. The grain was graded into different qualities in gigantic steam-powered elevators before being transported by ship or train to the West Coast and further over the Oceans. The animals were slaughtered in the huge slaughterhouses equipped with ingenious dis-assembly lines, and the meat was then transported in refrigerated railroad cars to the cities in the East. The lumber was sorted and dried in vast lumberyards, and most of the lumber was then sold to the farmers on the prairies, who needed a lot of timber for their houses, barns and fences.

In 1848 the first telegraph line reached Chicago. The combination of the vast volumes of grain and meat passing through the city and the access via telegraph to information about market conditions all over the world meant that the great hall of the Board of Trade became the world's leading food market in the 1860s. Thus, like their colleagues in Amsterdam before them, the leading merchants of Chicago became the avant-garde of a logistical revolution, in this case based on the railway and the telegraph. They grasped the potential of these infrasystems, and succeeded

[31]Cronon 1992.

in using them to gain control over the exploitation of the enormous natural riches of the prairies and the forests. They turned Chicago into Nature's Metropolis; to use a term coined by the historian William Cronon.

Pedestrian Cities

So far I have mainly discussed the role of what can be called *external* infrasystems in relation to cities, that is transportation and communication networks connecting cities with their hinterland and with distant markets. What about the role of the *internal* infrasystems within city borders? Up to the mid-nineteenth century, the developments of external infrasystems were most vital to the development of settlement patterns, but in the past hundred and fifty years the ability to develop internal infrasystems has been increasingly important to the development of such patterns.

In the nineteenth century a very fast urbanization took place in the Western world. It was spurred by the building of railways and telegraph lines, which provided a radical improvement in transport and communications, but only for cities and towns, not for the countryside. There were also other new infrasystems such as gasworks for lighting, which were available only in cities. Previously the availability of different natural resources had often been decisive in the location of industrial activities, but increasingly the availability of man-made infrasystems became decisive. Most of these systems were only accessible in cities.[32]

The cities of the nineteenth century were pedestrian cities. There were no efficient systems for transporting people within the cities, and thus most people had to walk to their work. This meant that factories and dwellings could not be too far apart, and larger cities in particular became very densely built and overcrowded. This implied a number of big threats. Sanitary and environmental conditions became more and more miserable and mortality rates increased. Big fires and popular riots were other dangerous threats. Finally the streets became more and more congested, making the transportation of goods and people more difficult and time-consuming, and this was a threat to the commercial activities in the cities.

A number of new infrasystems were developed to cope with these problems. Many of these systems implied a specialization and separation of different kinds of transport. In the first phase, a number of new pipe networks were built underground, below the city streets including gas, water and sewage systems. The latter two were especially important. They, in fact, addressed all the pressing problems mentioned above. When most inhabitants in a city had access to clean water and effective sewage, the mortality sank dramatically. Furthermore, fire hydrants improved fire

[32]Many examples are analyzed in Tarr and Dupuy 1988.

fighting and could also be used to suppress riots. And finally, the water and sewage pipes saved the streets and staircases from a lot of bulky transportation.[33]

The congestion of the streets was, however, not solved by these new underground networks and in big cities like London and Paris the situation became particularly chaotic in the middle of the nineteenth century. Different strategies were developed in the two cities for solving the problem. In the London region, political power was distributed over more than 300 different bodies, which made concerted action to cope with the problems very difficult. The building of an underground railway, the Metropolitan Line, linking all the main-line railway termini in the city, was by some groups seen as a radical way of improving passenger traffic. However, the construction of this line was very complicated and met much opposition, and it took twenty years to complete it. Furthermore, the smoke from the steam engines made journeys on this underground railway very unpleasant. It was not until the turn of the century after the introduction of electric traction, that the underground became a really viable alternative. In Paris, a more radical approach was pursued in the 1850s by the new, energetic prefect of the city, Georges Eugène Haussmann. He was appointed to his post by Napoleon III, and had a strong backing from the emperor, which gave him a very powerful position. Haussmann carried through a drastic reconfiguration of the entire city. Lots of old houses were torn down to give way to new broad boulevards, which provided fresh air and plenty of space for pedestrians and coaches. In addition, the boulevards made it much more difficult to organize riots and revolts such as the city had experienced in 1830 and 1848.[34]

We can talk of an intra-urban logistical revolution taking place in most major cities in the western world in the second half of the nineteenth century, and appearing in different forms, as I have indicated with the examples of London and Paris. But in almost all cases it involved the introduction of a whole array of new infrasystems providing efficient means for transportation of people, goods and information within the city borders. These systems made use not only of the space underneath the streets (gas, water and sewage, and subways), but also above the roofs (telephony and electricity), making possible an increase of the population and an intensified use of the urban space. It was the logistical revolution of the dense city.

Urban Sprawl

Around the turn of the century new infrasystems helped to break the old boundaries of cities, and a new form of urban expansion began. Electric trams and trains made it possible to commute from suburbs to the city-centre, and the access to telephone

[33] Jonsson et al., 2000.
[34] Hall 1998.

and electricity in the suburbs facilitated this exodus. The well-to-do families were the first to move to the new green and healthy suburbs, and the middle strata soon followed them. The poorest often remained in downtown housing areas, which turned into slums.[35]

The first wave out of the cities was organized originally around tram and railway lines and somewhat later also along bus routes. There was a certain concentration also in the suburban areas, because people had to walk or bike to the nearest tram- or bus-station. Thus, distinct settlement patterns evolved characterized by nodes along the new transportation lines. Not only dwellings were built in the suburbs; more and more factories, offices and shops were also located there, reachable by public transportation. But the different categories of activities were located in separate places, according to a new emerging planning ideal of functional separation.

The exodus out of the dense cities can be called the logistical revolution of urban sprawl. There have been two phases of this revolution in the past century. The first wave was centred on public transportation systems, and led to distinct suburban concentrations around the cities. The second phase centred around the motor car, which made it possible to live further away not only from the city but also from suburban centres, and this contributed to urban sprawl over vast areas. More roads had to be built to cope with growing traffic volumes, and new shopping centres and work places emerged along these roads accessible only to households with cars. These developments stimulated more and more people to go by car, and in many urban regions this has led to a vicious circle of increasing traffic congestion.

However, there are big differences among cities regarding their degree of sprawl and their dependence on the car. One important factor seems to be the age of the city. In general, young cities (which are more common in North America and Australia than in Europe) have a high degree of sprawl because they have experienced most of their growth after the introduction of the car. Los Angeles is the archetype of this development. Already in the mid 1920s almost every household had a car, and at this time the city politicians and planners decided to develop a new network of freeways instead of trying to preserve the existing streetcar and light rail systems. This strategy led to a decline of the old downtown area and a very high degree of sprawl and an almost total dependence on motorcars.[36]

In contrast to cities like Los Angeles, most European cities already had a rather large and densely built downtown area when the car was introduced, which made a whole-hearted car strategy almost impossible due to lack of space for roads. Many European cities have instead developed a conscious strategy of supporting and developing public transport. A good example is Stockholm, which is a city largely

[35]Ibid.
[36]Ibid.

located on islands, and thus dependent on a limited number of bridges. This made transportation a critical issue in the mid twentieth century as more and more households wanted to move out of the inner city. In the decades after World War II, Stockholm built an elaborate underground system under the downtown area which later spread far out of the city. The construction of the underground lines was closely connected to the building of a whole series of new suburbs along these lines, like beads on a necklace. These suburbs were carefully planned with a commercial centre, schools and high-rise buildings close to the station, and with lower dwellings further away from the station, but still within walking distance. Furthermore there were plenty of forests and recreation areas close to each suburb. This co-ordinated suburban development along underground lines was made possible through a rather high degree of mutual understanding among the leading politicians from different political parties and also among politicians and real estate owners. As a result of this strategy, Stockholm achieved a very high percentage of public transport use compared to many other cities of a similar size.[37]

With this short historical exposé of logistical revolutions in the past, I have primarily wanted to illustrate that the combined effect of intertwined infrasystems on the spatial distribution of commerce, industry and settlements can be very powerful. I want to emphasize that there has not been one uni-directional development. There are often many different ways in which infrasystems can be combined. A logistical revolution should thus be seen as a formative phase, when new infrasystems offer new opportunities and when new combinations of infrasystems are possible to achieve. Cities have differed in their response to these opportunities. Above, I have primarily discussed cities, in which the power elite has been particularly skilful in grasping the opportunities offered by new infrasystems and has developed clear strategies and plans for how to implement them at an early stage. Many other cities and urban regions have had a more piecemeal approach, and have rather followed the examples of others.

Once an urban region has chosen to develop and combine new infrasystems in a certain way, a process of embedding starts making future change difficult. A distinction can be made between *structuring* and *adaptive* infrasystems. The dominant transportation system at a given time often has a stronger structuring effect than other systems, which rather adapt to the new patterns shaped by the transport systems. Therefore the choices and the designs of transportation systems have often been particularly important for the long-term development of urban regions. Another way to phrase this is to say that rails and roads have a strong tendency to almost literally contribute to the phenomenon of path-dependency.[38]

---

[37]Johansson 1987; Hall 1998.
[38]David 1988.

What Can We Learn from History?

'In sum it is difficult to change the direction of large electric power systems – and perhaps that of large socio-technical systems in general – but such systems are not autonomous. Those who seek to control and direct them must acknowledge the fact that systems are evolving cultural artefacts rather than isolated technologies. As cultural artefacts, they reflect the past as well as the present. Attempting to reform technology without systematically taking into account the shaping context and the intricacies of internal dynamics may well be futile. If only the technical components of systems are changed, they snap back into their earlier shape like charged particles in a strong electromagnetic field. The field must also be attended to: values may need to be changed, institutions reformed, or legislation recast.'[39]

This quotation is from the concluding paragraph of Thomas P. Hughes' book *Networks of power*. Hughes here underlines what I think is the most important general lesson from history. Infrasystems are social constructions; they do not develop in some autonomous, uncontrollable way, even if it sometimes may seem so due to the 'momentum' that many of them have acquired. It *is* possible to redirect systems, but this presupposes that an alliance of interests can be formed that is powerful and persistent, and whose actions are based on an understanding of the socio-technical nature of infrasystems.

In this concluding section of the article, I will try to draw some conclusions and lessons from history that may guide those who want to contribute to the redirection of infrasystems in a sustainable direction. I want to underline that I am deliberately vague when talking about a redirection of infrasystems 'in a sustainable direction', and what this should imply more specifically. In my opinion it is not the task of a historian to define what kind of changes that are desirable for the future. But a historian can assist those who want to bring about processes of change, by trying to provide lessons about the general character of such processes.

Infrasystems Are Socio-Technical Systems

A first lesson is that infrasystems are socio-technical systems, in which the institutional frameworks and the system culture are as important as the technical components. In the public debate, there is generally a lack of understanding of the importance of these 'soft' parts of infrasystems and a strong belief in 'technical fixes'. However, a prerequisite for achieving lasting changes is that the system culture and the institutional conditions are altered. I have argued that the establishment of infrasystems has often been dependent on a crucial institutional innovation,

[39]Hughes 1983, p. 465.

which made it possible to overcome the initial uncertainty by distributing the huge capital costs for building facilities and networks among many users. In fact, the communal use and the public accessibility (to all that are willing to pay for the services) is the very essence of infrasystems.

The institutional shaping of an infrasystem can be seen as the result of an encounter in the past between technology and society. For this reason the institutional frameworks for infrasystems have differed considerably both among systems and among countries and regions. The frameworks of infrasystems have often been rather stable over time, and they contain a heavy legacy. This makes it important to learn about their history. They were largely shaped long ago by people, who conceived a number of problems to be overcome and opportunities to grasp. It is also important to remember that they have often been shaped in societies that were not democratic, but in which a small elite had the political power and used it to promote their own ends. In general, these frameworks have primarily been shaped to facilitate the expansion of infrasystems, simply because the positive effects of the systems were much more obvious than the negative ones, in particular for the wealthy. This urge for expansion is in many cases deeply imbedded in the system culture permeating the organizations owning and operating the systems.

When trying to redirect infrasystems in a sustainable direction, it is thus crucial to make changes in their institutional frameworks and in their system culture in such a way that strong incentives are created for finding solutions that are environmentally benign. There is almost always strong opposition within organizations towards changes of this kind. A prerequisite for accomplishing them is therefore that a broad alliance can be formed, including people from all spheres of society, including political parties, environmental organizations, industry, public authorities, trade unions and universities. There seem to be circumstances that can facilitate such changes in the coming years; it can be argued that we are at present in a formative phase. The reason is that rather far-reaching changes in the institutional frameworks of infrasystems have taken place in many countries in the past ten to fifteen years under the heading of 'deregulation'.[40] The primary aim has been to increase the economic efficiency of infrasystems by stimulating competition among many system operators and by facilitating cross-border traffic. These reforms are in an early stage and have not become firmly embedded as yet. At the same time the importance of environmental considerations is being understood in ever wider circles. These two processes may together provide a window of opportunity for implementing institutional changes, which will make environmental considerations imperative in the development and operation of infrasystems.

---

[40]Re-regulation is in fact a more adequate term. That is; regulation has changed, not diminished.

Dynamics of Infrasystems

A second lesson pertains to the dynamics of infrasystems, and here one main message is that it is important to realize that infrasystems go through phases with different conditions. Policies for change have to take this into account.

For infrasystems in an early stage the major challenge is to overcome uncertainties and to find a first market where it can be established. In fact, most attempts to establish new infrasystems fail. The establishment of a new, environmentally benign system may therefore need substantial support from public authorities to overcome initial uncertainties.

When a system has been successfully established in a first market, it may reach a phase in which strong economic and social forces for expansion create a fast spiral of growth. The overriding concern for the system operators and politicians during such periods is generally to ensure an expansion of the system that is fast enough to meet the growing demand. Long-term effects of the systems on the environment and in other aspects tend to become secondary, precisely in the period when most of the long-lasting hardware is installed.

A phase of fast expansion often brings about a system culture in which future growth is taken for granted, and it therefore often takes a long time for the managers of infrasystems to anticipate a stagnation or decrease in demand of their services. This may lead to the creation of a substantial overcapacity, as has been the case for energy- and water supply systems in many industrialized countries in the past decades. Such an overcapacity diminishes incentives for efficient use of services.

Another important issue, in relation to the dynamics of infrasystems, is how environmental problems can be given a higher priority when developing new components or sub-systems. Particularly in the expansion phase of infrasystems obstacles or 'reverse salients' have appeared threatening the further expansion. Typically such 'reverse salients' have pertained to the increase in scale of components or sub-systems, and the incentives to overcome them have been very strong. Leading engineers and scientists have often been engaged in such efforts, and at a number of times, radical new components or entirely new system designs have been the outcome of such efforts. A key question is then how the environmental effects of infrasystems can become 'reverse salients', attracting the interest of leading engineers and scientists.

Since the 1970s, there has been a growing environmental movement in many industrialized countries, and this has spurred politicians to promote environmentally benign technologies, not least in energy and transportation systems. One policy has been to 'put a price on the environment' by introducing taxes or fees on scarce resources or on pollution. This has certainly stimulated the development of system components, which are more resource-efficient or have lower emissions, but it has seldom led to the sense of urgency that is necessary to create a reverse salient. Another policy has been to introduce compulsory environmental standards or to forbid certain kinds of dangerous substances. These kinds of policies have been rather successful in some cases, for example diminishing the use of CFC-gases and introducing lead-free petrol. However, industrial interests have often been able to prevent or at least moderate such legislation. It is the latter kind of policy, if pursued in a consistent and determined way that has the potential to make 'reverse salients' of environmental problems.

The concept 'functional equivalent' is important in this context. For example, in the case of CFC-gases, alternative technical solutions – functional equivalents to the CFC-gases – existed, which made it possible for legislators to pursue tough policies, demanding an abandonment of CFC-gases. It is often of critical importance to be able to show that a functional equivalent to a polluting component or system exists, even if it is expensive to make a substitution. Research to develop such functional equivalents may pose a threat to the dominant actors within an infrasystem, and therefore public funding of such research is essential.[41]

There is also another problem. Developing new environmentally benign technologies is not enough. They also have to be broadly adopted. The car industry illustrates this dilemma. While new cars are being developed with ever-lower fuel consumption and emissions, an increasing number of customers prefer to buy bigger and bigger cars (vans, jeeps, SUVs etc.) and they seem to be rather insensitive to costs.

Interplay among Infrasystems

A third lesson is that interplay among infrasystems can be of crucial importance. On the one hand competition among systems fulfilling the same function has been a major factor in the development of many infrasystems both as a stimulus for technical and economical improvements and as a cause for decline. On the other hand infrasystems have often played a complementary role to each other producing synergistic effects, as illustrated by the railway and the telegraph. This kind of complementary interplay is increasing rapidly. In particular, modern information and communication technologies are becoming more and more essential for the management and operation of all kinds of infrasystems. This development is double-edged. On the one hand it provides a potential for improving co-ordination within and among systems, not least in the transport sector. For example, most major freight companies have introduced new information systems in the past decade enabling a substantial increase in the load factor of lorries. There is, however,

[41]Kaijser et al., 1988.

another side to this coin. A growing interwovenness of infrasystems will also lead to an increasing complexity and vulnerability. A breakdown in one system may get almost instant repercussions in many other systems. This vulnerability was clearly illustrated by the extremely costly preparations for avoiding breakdowns on the New Years night of the new Millennium.

Another kind of interplay among systems is the joint use of networks. Building new networks or rebuilding existing ones is always very expensive, and in particular in dense urban areas.[42] This makes existing networks a very valuable asset, and they can play a crucial role for introducing new systems. For example, Internet services can be brought to a new customer in a number of ways: either by using existing telephone lines, electricity lines, TV-cables or radio transmission or by installing a new optic fibre. Using existing networks does not provide the same capacity as optic fibres, but is much cheaper and can be implemented much quicker. In the case of the Internet, the owners of the 'host networks' are happy to provide capacity to Internet traffic as this does not threaten their traditional services and increases their incomes.

In many other cases, access to existing networks will be contested. One example concerns the distribution of electro-magnetic frequencies, which can function as networks for many different kinds of systems. Broadcasting companies were among the first players on the scene, and they divided many of the most attractive frequency ranges among themselves from the 1920s and onwards. But in the past decades many new kind of systems have been developed which make claims on frequencies, and this has led to hard negotiations about reallocation of frequencies. Another example concerns urban roads and streets, which have been used by many different modes of transport in the past. At times there has been intense competition among systems for street space. In the mid twentieth century this competition was aggravated by an increase in the number of cars, and in many cities streetcar systems were abandoned in order to give more room to cars and buses.[43] Since then, cars have achieved a dominant and privileged position on the streets. This privileged position of the car will most probably be renegotiated in coming decades, in order to provide room for more environmentally benign and space efficient transport systems. However, the influential motorcar lobby organizations will probably fight hard to prevent such a development. In the past years they have strongly opposed the introduction of road pricing, which is probably the most effective way to achieve a more efficient use of scarce road space.

[42]The huge cost of rebuilding existing networks is illustrated by the so-called 'Big Dig' in Boston. The replacement of an elevated motorway cutting right through the downtown with a tunnel of a few kilometers length will cost at least 20 billion USD.
[43]Ekman 2000.

Infrasystems and Settlement Patterns

A fourth lesson is that there is a strong interrelation between infrasystems and settlement patterns, in particular in urban regions. Most infrasystems have first been built between or within urban areas, and the access to these systems became an important competitive advantage for cities and towns in relation to rural areas and spurred a fast urbanization. Furthermore, the development of new systems, and in particular transportation systems, influenced the status among cities. Those that were able to attract or build such systems at an early stage often prospered and advanced in the urban hierarchy, while those that were late often sank. Until the late nineteenth century, all cities were basically pedestrian cities, as there were no efficient systems for transporting people within them. Around the turn of the century, the introduction of electric trams, commuter trains and later on also buses and cars enabled those who could afford it to move out to new suburbs. The ways in which these transportation systems were built, had a strong and long-lasting impact on urban settlement patterns. These systems thus had a structure-shaping character.

A conclusion of this is that it is absolutely essential to make assessments of the long-term structural effect when the construction of new airports, streetcars, highways, bridges or other large infrasystem projects are being considered and to let such assessments have a key role for the decision. This is, unfortunately, seldom the case. Mostly, investment decisions are largely based on assessments of the effects on traffic flows once the project is completed. The mega-cities in the third world are the most rapidly growing cities in the world, and it is particularly important to develop and implement policies for infrasystem development in these places, which focus on long-term consequences for settlement patterns.

Historically, transportation systems have had the strongest structure-shaping effects. However, in the future information technologies may partly take over this role. In the industrialized world a large building stock has already been built, and a redistribution of activities within the existing housing stock may become more important then the construction of new buildings for lifestyles and travel patterns of urban populations. New information and communication technologies are an important factor in this context, as they provide new opportunities to carry out many kinds of activities. We seem to be in a formative phase, as these technologies can be used for many purposes and in different ways. One scenario is the revival of more integrated towns, with shorter daily travel needs as a result. Another scenario is a further urban sprawl. A crucial question for a societal development in a sustainable direction is what kind of policies will prevent the latter scenario.

I hope this chapter may give some insights to those who want to redirect infrasystems in a sustainable direction. Let me repeat the main message. It *is* possible to redirect systems and it is important to do so, but it is difficult. It presupposes that an alliance of interests can be formed that is powerful and persistent, and whose actions are based on an understanding of the socio-technical nature of infrasystems.

References

Andersson, Å. E., and Strömquist, U., (1988), *K-Samhällets framtid*, Stockholm.

Andersson-Skog, L., and Krantz, O., (1999), *Institutions in the transport and communication industries. State and private actors in the making of institutional patterns, 1850–1990*, Canton, Ma.

Blomkvist, P., and Kaijser. A., (eds) (1998), *Den konstruerade världen. Tekniska system i historiskt perspektiv*, Stockholm.

Braudel, F., (1979) (Engl. transl 1992), *Civilization and Capitalism 15th to 18th Century, Volume III, The Perspective of the World*, Berkeley.

Braun, I., and Bernward, J., (eds) (1994), *Technik ohne Grenzen*, Frankfurt am Main.

Coutard, O., (ed.) (1999), *Governing large technical systems*, New York.

Cronon, W., (1992), *Nature's Metropolis: Chicago and the Great West*, New York.

David, P., (1988), *Path-Dependence: Putting the Past into the Future*, Stanford.

Dobbins, F., (1994), *Forging Industrial Policy. The United States, Britain and France in the Railway Age*, Cambridge, MA.

Egyedi, T., (1996), *Standardisation of the container. The development of a gateway technology in the system of cargo transportation*, Stockholm.

Ekman, T., (2000), *Kampen om gatan: Avvecklingen av spårvägstrafiken i Stockholms innerstad, 1920–1967*, Stockholm.

Elton, A., (1958), 'Gas for Light and Heat', in C. Singer (ed.) *A History of Technology*, Vol IV, Oxford.

Fridlund, M., (1999), *Den gemensamma utvecklingen: Staten, storföretaget och samarbetet kring den svenska elkrafttekniken*, Stockholm.

Grübler, A., (1990), *The Rise and Fall of Infrastructures. Dynamics of Evolution and Technological Change in Transport*, Heidelberg.

Hagberg, J.-E., (1986), *Tekniken i kvinnornas händer. Hushållsarbete och hushållsteknik under tjugo- och trettiotalen*, Malmö.

Hall, P., (1998), *Cities in Civilization*, New York.

Heimburger, H., (1931), *Svenska telegrafverket: historisk framställning, Band I: Det statliga telefonväsendet, 1881–1902*, Göteborg.

Helgesson, C.-F., (1999), *Making a natural monopoly*, Stockholm.

Hirsh, R., (1989), *Technology and Transformation in the American Electric Utility Industry*, Cambridge.

Hughes, T. P., (1983), *Networks of power. Electrification in Western Society 1880–1930*, Baltimore.

Hughes, T. P. (1987), 'The evolution of large technical systems', in W. Bijker, T. P. Hughes, T. Pinch (eds) *The social construction of technological systems. New directions in the sociology and history of technology*, Cambridge, Ma, pp. 51–56.

Hughes, T. P., (1992), 'The Dynamics of Technological Change: Salients, Critical Problems, and Industrial Revolutions', in G. Dosi, R. Gianetti and P. A. Toninelli (eds) *Technology and Enterprise in a Historical Perspective*, Oxford.

Hård, M., and Olsson, S.-O., (1995) 'Enforced Marriage: How District Heating and Electricity Systems Have Been Combined', in A. Kaijser and M. Hedin (eds) *Nordic Energy Systems: Historical Perspectives and Current Issues*, Canton.

Johansson, I., (1987), *Storstockholms bebyggelsehistoria. Markpolitik, planering och byggande under sju sekler*, Stockholm.

Jonsson, D., (2000), 'Sustainable Infrasystem Synergies – A Conceptual Framework', *Journal of Urban Technology*, Vol 7, Nr 3, 81–104.

Jonsson, D., Gullberg, A., Jungmar, M., Kaijser, A., Steen, P., (2000), *Infrasystemens dynamik – om sociotekniska förändringsprocesser och hållbarhet*, Stockholm.

Kaijser, A., (1986), *Stadens ljus: Etableringen av de första svenska gasverken*, Malmö.

Kaijser, A., (1993), 'Lighting and Cooking: Competing Energy Systems in Sweden, 1880–1960', in W., Aspray (ed.) *Technological Competitiveness: Contemporary and Historical Perspectives on the Electrical, Electronics, and Computer Industries*, Piscataway, NJ, pp. 195–207.

Kaijser, A., (1994), *I fädrens spår. Den svenska infrastrukturens historiska utveckling och framtidda utmaningar*, Stockholm.

Kaijser, A., (1999), 'The helping hand. In search of a Swedish institutional regime for infrastructural systems', in L. Andersson-Skog and O. Kranz (ed.) *Institutions in the transport and communications industries*, Canton, Ma, pp. 223–44.

Kaijser, A., Mogren, A., and Steen, P., (1988), *Att ändra riktning: Villkor för ny energiteknik*, Stockholm.

LaPorte, T., (ed.) (1991), *Social responses to large technical systems. Control or anticipation*, Dordrecht.

Lilley, S., (1973), 'Technological Progress and the Industrial Revolution 1700–1914', in C. Cippola (ed.) *The Fontana Economic History of Europe Vol 3*, London.

Mayntz, R., and Hughes, T., (eds) (1988), *The development of large technical systems*, Frankfurt.

Summerton, J., (1992), *District heating comes to town. The social shaping of an energy system*, Linköping.

Summerton, J., (ed.) (1994), *Changing large technical systems*, Boulder.

Tarr, J., and Dupuy, G., (1988), *Technology and the Rise of the Networked City in Europe and America*, Philadelphia.

de Vries, J., (1981), *Barges and capitalism. Passenger transportation in the Dutch Economy, 1632–1839*, Utrecht.

de Vries, Jan, and van der Woude, A., (1995), *Nederland 1500–1815. De eerste ronde van de moderne economische groei*, Amsterdam.

PART II

COMMENTARY

A Dialogue Concerning the Usefulness of the Social Sciences

Bengt Hansson

General format:
The Minister for the Environment (ME) is worried about the slow progress in changing people's environmental behaviour and has decided to call upon a social scientist (SS) to sort things out

ME: I am worried. I am much concerned about the environment, and so is the whole country. I also know very well that it is really the expectations, life styles and habits of the citizens that ultimately cause the environmental problems, even if the problems are often described in terms of their effects on air, water, radiation, biological variation and that sort of thing. But it seems so much simpler to go for these tangible natural effects than to address the real social causes. We know too little and things happen too fast. I mean, new trends in life styles, food habits, travel patterns, and so on pop up all the time, and we don't notice their environmental effects until after a while. And then we start raising awareness, launching information campaigns, creating environment friendly alternatives, legislating, and so on. All this takes time, and usually more time than it takes for the trend to transform itself into a new trend. New problems turn up faster than we solve the old ones.

So, let me be frank – what's the point with doing more social science research? Don't we know enough? Isn't it high time to act instead? There are such a lot of studies of this and that, but no really hard and useful results. Everything is so context dependent and of so limited applicability, and every piece of research seems to result in the conclusion that two new pieces are urgently needed. Even if we could afford such proliferation, it seems to me that it would do more harm than good because it would give as such a fragmented picture.

SS: Well, well. You seem to have made up your mind already to do something. What card is it that you have up your sleeve?

ME: Oh, no, I didn't say I had a simple solution, not even that I had any definite ideas about what to do. Don't misunderstand me, I am not simplistic about complex problems. And I have absolutely nothing against doing social science *per se*. All I do is raise the question whether it is likely that it will be any good for *me*. Should I wait indefinitely for firm results, or should I forget about social science and go ahead (and maybe hope to reap some windfall later on)?

SS: OK, let me see if I have understood you correctly. Your problem is this: you think that it is essential to influence people to behave in a more environment friendly way, and you want to know if there are, or soon will be, results from the social sciences that help you to do so more efficiently than you could have done otherwise. Is that the question you have hired me to answer?

ME: Yes, right, that's it. And what's your answer? And by the way – were you not right now falling into that old trap of the social sciences: to put simple things in pompous language, claiming spurious precision?

SS: I'm glad you confirmed my interpretation. My answer to the question is yes, and the reason I am here is to prove it. And to furthermore prove the value of precision, I would like to point out what is *not* implied by my positive answer: I have *not* promised that the social sciences will be able to give you definite recommendations what to do. They *will*, however, be able to help you see what *sort* of thing you should try to do, what is feasible or not feasible, which measures will be mutually reinforcing or the opposite, which effects will be lasting and which will not, and similar things. But you were aware of that distinction all the time, weren't you?

ME: Eh, of course. Let's get down to business.

SS: Meaning what?

ME: Well, since I am not supposed to ask you for definite recommendations, I really don't know how to start. But I do know that the results in the social sciences are surrounded by too much vagueness and uncertainty to turn easily into recommendations, so it is not enough if you just present me with results. I hope you are not going to say that you will just present me with the data and that it is my own responsibility to draw the right conclusions. Then I would be the social scientist myself, and I wouldn't need you. Come up with something better if you want to keep your job!

SS: Actually, I *was* going to say that much of the responsibility for the conclusions will in fact rest with you. But not because you will play the role of the social scientist; there is a different role for you, with different responsibilities. Oh, I see you shrink back a little. But I think this is important – it concerns in fact the very heart of the applicability of the social sciences – so I hope you will excuse me if I take a few minutes to spell it out. OK?

ME: OK.

SS: You just mentioned the 'vagueness and uncertainty' of social science results. Why? Probably because you were thinking of studies where a number of people, most likely drawn from several categories, have been interviewed about certain issues and their answers tabulated and interpreted. The validity of the scientist's conclusion depends on the representativeness of the sample, the neutrality of the interviewer and the interview situation, the suitability of the phrasing of the questions and such things. Other types of studies have other types of preconditions. The social scientist tries to safeguard against spurious results by using statistical significance tests, standardized interview procedures, independent evaluators and other such precautions. Every single study is tightened up as much as possible, but yet the result should properly be stated only as '*if* the preconditions hold, then so-and-so'.

This is *one* sort of uncertainty, within a single study. To increase precision in this respect is the sole responsibility of the author. But if you wish to apply a result, you will face another kind of uncertainty. Say that you have learned from a real-life experiment that a certain type of refuse bin has significantly increased refuse separation in a certain residential area. Should you prescribe the use of that type of bin for the whole country? That depends: Will the result still hold for other residential areas? For families in other phases of their life course? Also in the winter? After the initial fascination for the gadget has faded? The general question is: How much can we generalize the result? What is its range of applicability?

This is *another* type of uncertainty. It requires that you can judge what is *essential* in a situation and what is only *accidental*, or, expressed differently, which are the true causes of the result. Was the bin in question successful because of its own inherent qualities or because it fitted well with time- and context-sensitive characteristics of the residential area? Suppose you can firmly establish that it was because of its own inherent qualities. Then the research result could immediately be transformed into a definite solution: a particular kind of refuse bin. But if the alternative explanation were in fact correct, the result would be, at best, a *method* for adapting bins to local and current circumstances. In fact, I think that most results in the social sciences are like this, and that is why decision makers have to be reminded that they should not expect the same from the social as from the natural sciences – less gadgets and definite blueprints and more understanding of interrelationships.

To separate the essential from the accidental is thus of outmost importance for the application of a research result, but it is seldom achieved within a single study. It requires comparisons of many types of studies, carried out under varying circumstances, and it is not nearly as data dependent as most single studies are. Rather, it consists in an ability to place the many studies that are produced in various disciplines into a common framework to see how they all hang together. It takes both some basic common sense and a good acquaintance with methodological matters to distinguish between real and apparent contradictions, bring out interdependencies and identify white spots on our knowledge map.

ME: OK, OK, I agree with all that, but why should *I* be responsible for doing it? Isn't that a matter for the scientists too?

SS: Yes, it is. Or rather for science as a whole than for individual scientists. We urgently need critical surveys of high quality, and there are some problems getting the best people to do that. But it is *also* your responsibility. It is part of the responsibility of any politician – or, more generally, of anyone responsible for implementing policies based on scientific results – to take *all* relevant aspects into consideration and to evaluate both the extent and the quality of our knowledge base. Somebody simply has to make the decision that we now know enough to go ahead and do something, or else that we need to investigate this or that before we do so. And shouldn't that somebody be you? If you want a fancy word, it is 'receiver competence'.

ME: So what you *really* want to say is that your task is not to feed me with either recommendations or scientific results, but to educate me to be some kind of superior court for scientific results? OK, I am willing to set aside half an hour a day to do that.

SS: Do I hear a slight tone of sarcasm? OK, I got carried away and sounded more pompous than I really intended. But it is important that someone takes a step back and puts things in perspective. To increase precision in the detailed work takes one kind of mind, to assess the balances in the whole picture takes another. They are not related as superior or inferior, but as complementary and parallel. What I want you to be is a parascientist, not a superscientist. And of course not you personally, but yet someone with a mandate to act in your name.

ME: Yes, yes, I wanted to do exactly that – put things in perspective – concerning the place of science among my own duties. But of course I see what you mean. However, time is up for today. Next time, let's be more focused. It is up to you to select a well-defined topic, OK?

SS: OK.

Next week:
Environment friendly behaviour is more common among some people than among others. What can we learn from that?

SS: It would be good if I knew more about your preconceptions about what research in the social sciences is like. Let me ask a simple test question. Suppose you have to deal with a certain environmental problem and that you are convinced that it is people's behaviour in a broad sense – the way they choose to live, eat, travel, shop – that is the ultimate cause of the problem and therefore needs to be changed. What type of social science study would you look for in order to get useful information for that sort of problem?

ME: Well, in my way of thinking it is always better to try to reinforce tendencies that already exist than to try to force an entirely new picture on people. So I would look to investigations that categorize people in relevant ways and study which categories of people are already on the right track.

SS: What categories are you thinking of?

ME: Oh, I suppose sex, age, family situation, income, education, occupation, place of living, political views – that sort of thing.

SS: What you might call traditional sociological variables, then?

ME: I suppose so.

SS: Have you any clear expectations what such studies will show?

ME: Women probably show more concern for the environment than men, the young and the old more than the middle-aged, the better educated and higher paid more than the poor and less educated, people on the countryside more than city dwellers. Or is that wrong? Are you testing my prejudices?

SS: In a way, yes. What you say is certainly very conventional. But I will give you a second chance to venture a more sophisticated opinion. The leading question is: were you in fact thinking about people's *concerns*, as you expressly said, or rather about their actual impact on the environmental load?

ME: Well, I am living in a world of voiced opinions and I get rather little feedback about the actual impact of individuals, so I guess that my mind has been impressed more by words than by deeds. But now that you mention it, there was something about age that made me uncertain. What did I say? That the young and the old showed more concern than the middle-aged? Yes, and when I said 'the young' I was certainly impressed by their idealistic concern for the environment, but for 'the old' I rather thought about their less extravagant style of life, meaning their actual impact. Prejudices of course, but different prejudices for different groups. Does it matter? Is there a big difference between what people say and what they do?

SS: It is a bit embarrassing, but we know rather little about that. Most studies focus on one of a small number of self-reported token actions, conventionally considered to be examples of environment friendly behaviour, such as using public transportation, separating refuse, buying eco-labeled detergents, or using low-energy lamps, and we really don't know how well this sample of actions represents the overall ease on the environmental load. In other words, most of our studies cannot distinguish between actions made for their token value and actions made for their real effects.

ME: Isn't that just a matter of information? If people show good will in some cases but fail to do the right thing in others, isn't that just because they don't understand the consequences?

SS: I don't think so, not always anyway. Information is a very tricky business and deserves a full discussion in its own right. But the token value of an action should not be easily dismissed – it serves as a means for forming identities and for earning commendation. For example, as Anna-Lisa Lindén has shown, the way to make people buy 'green' electricity is to enhance the token value of the action. Another example is provided by people who have chosen to live in so-called eco-villages. They have committed themselves to certain kinds of environment friendly behaviour, and they usually stick to it rather conscientiously. But they do not seem to differ much from the population in general in terms of their overall contribution to the environment load. This observation is due to Mona Mårtensson in one of her studies of everyday practice and the environment, one of the few that has a broad enough perspective to allow this kind of observation. But this is somewhat of an exception; the things we know most about are verbal reports of attitudes and of certain token actions.

ME: Given that, how right was I in my guesses?

SS: So-so. You were probably right in saying that women are more concerned than men, at least in certain respects, but you were wrong in your beliefs about the well educated and better paid. Perhaps you thought that the educated understand the complexities of environmental effects better, or that the well-to-do can afford more long-term considerations, but there is little evidence that points in that direction.

But I would like to challenge you, not about the correctness, but about the relevance of your guesses, or rather your hunches, as I think they were. Suppose you were absolutely and emphatically right about women, that all studies showed that women were much more concerned about the environment in both words and deeds. So, what? How would that affect your policy?

ME: Well, I couldn't turn the rest of the population into women, obviously. Perhaps I should reason like this: the women take care of themselves, so I should focus on the men and try to influence men's behaviour specifically. How about that?

SS: Not a bad thought. There may be some fine-tuning needed about the comparison of the marginal efforts required to influence men and women respectively, but the idea is basically sound. Not, however, sufficient. For how do you go about it? The mere fact that men and women have different attitudes and put a load on the environment to different degrees doesn't tell you how to change it. You must have an idea of *why* there is a difference.

ME: No, of course, it is not likely that it is womanhood or manhood as such that lies behind the difference. And if it were, there wouldn't be much we could do about it. So it must be something else, associated with sex in a more contingent way. Perhaps upbringing, perhaps different household roles, perhaps society's expectations. And yes, these do change, even if the pace is slow. Is this what you meant by separating the essential from the accidental in a situation?

SS: Yes. Some differences *show up* as sex differences because you put together your tables that way. But the difference may be only accidentally connected with sex, and may depend essentially on some underlying factor, for example, different household roles. Incidentally, this is one of the good reasons, and a non-ideological one, why it is often better to speak about gender than sex. It reminds you that the connection with attitudes and behaviour is not fixed once and for all. Mona

Mårtensson and Ronny Pettersson speak about 'gender order', by which they mean 'historically construed patterns of relations between women and men and cultural ideas about this pattern', and they found it to be a contributing determinant for several types of behaviour and central to travelling by car.

ME: Good, that's something we try to change anyhow, so we may get some environmental benefits thrown into the bargain.

SS: Perhaps, but don't take it for granted that equalization of gender roles is necessarily good for the environment. It makes a difference if men adjust to women or the other way round.

ME: Of course. Now I will try to make a summary note of this week's talk. How about this: 'If a study in the social sciences is to be useful for me, it has to go beyond results describing what is immediately observable and discuss underlying causal patterns, because it is these patterns that determine what can be done.'

SS: Right, but there is another side of the coin. The very fact that underlying patterns *can* be changed means that they are context dependent and that brings us back to the question how much the result of a single study can be generalized. Remember the miraculous refuse bins? You must not presume that that which has worked in one setting will always work in all settings. So I suggest you make a note of last week's conclusion too: 'I should always try to look at several studies and compare them in order to get an idea of the range of applicability of the result.'

ME: Fine. This goes well; it's not too difficult – rather much common sense actually, once you've helped me to sort it out.

SS: Don't relax too soon. These have just been warming up sessions. Next week we will speak about age-related behaviour, and that will be a harder nut to crack.

Third session:
Interpreting age effects – cohorts and life course effects

SS: Now let's return to your hunch that older people put less of a load on the environment than younger ones. When you thought about it you mentioned their less extravagant style of life as a reason why this should be so. Is that right?

ME: Yes, isn't that a common impression? It may vary with socio-economic class and place of residence, of course, but generally speaking I think it is true. Or is there a more sophisticated truth in social science?

SS: No, I think you are right. The old form another category which is, to use your own words, on the right track, just like the women. But can you use that fact for action or prediction?

ME: I said before that I couldn't turn men into women, and now I cannot make the young ones age by just snapping my finger.

SS: Are there perhaps other means than finger-snapping?

ME: Are you joking? Or do you mean that they will age anyhow? But that will take a generation!

SS: True, and I understand that you are eager to find quick solutions. But you certainly need to control long-term developments too. I think that it is in fact all-important to do so, but I will save my arguments for later. But if we now think a generation ahead, what do you think will happen?

ME: Squarely, the present generation of elderly will have disappeared, the now middle-aged will have become the old ones, entering that age of more environmental friendliness, and the now young will have become middle-aged, bringing with them, I hope, the environmental enthusiasm of their youth. Hey, that means that things are working my way more or less automatically!

SS: Unfortunately not. To see why, let's begin with that first transition, from the now middle-aged to the future old. Why do think that they will become more environment friendly with age?

ME: But it has been established that older people are more environment friendly, hasn't it? I mean, on the average.

SS: Yes it has. But I hope you remember the note you made last week – look not only at the figures but seek the underlying causes!

ME: So I have to ask myself, why are older people more environment friendly? Well, that's common sense: because they have been raised under different conditions, when everything was scarcer. They had to learn to economize not only with money, but also with space, material things, and travel. You can say that it became a second nature to them to live a more restrained kind of life, and those habits have lingered on – everything on the average, of course.

SS: That sort of mechanism is certainly operating to some extent, but there may well be other reasons too. We'll come back to that. But first I wish to draw your attention to the fact that you are making two tacit assumptions here: first that people are particularly imprintable during their young years and that their attitudes will not change easily even if they experience quite different conditions later in life, and secondly that the times when the now old were young were marked by scarcity, at least relatively speaking. The second one is an easily ascertainable matter of fact, but the first one is a rather general and elusive assumption about human nature. Do you really believe in it?

ME: Haven't we all noticed how preoccupied old folks are with their childhood memories, also as a measuring stick for today's customs and values?

SS: We have also noticed that quite a few of them pick up modern ways of thinking about e.g. what to expect from the health care system. But let's leave that for the moment. If you are right you have identified what is called a *cohort effect*. A cohort in a population consists of all those who are born in a certain year, or during a short specified period. A cohort effect means that everyone in a cohort is affected similarly by something typical for their youth years, and that effect will then remain in the population, although it will move, as time goes by, to increasingly older strata of the population.

ME: Yes, we see that in many other cases than environment friendliness – eating habits, sexual morality, for example. But some things are more deeply entrenched than others, so perhaps we shouldn't be too quick to draw conclusions.

SS: No, but let's try to be sharper rather than vaguer. *If* the environment friendliness of the now old is really a cohort effect, what does that tell us about the old of the next generation, the now middle-aged?

ME: Eh, nothing immediately. It all depends on the characteristic features of *their* youth. Just because times some 60 years ago were marked by scarcity, it doesn't mean that times 30 years ago were the same. In fact, I think they were the opposite. And then the future old will be *less* environment friendly than the rest of the population. It seems that things are not working my way automatically after all.

SS: Only if we explain what we have found about the old as a cohort effect. But perhaps you can think of some other explanation?

ME: Oh, I have lots of prejudices about the old in general, and I see no reason why most of them shouldn't be right. So I would say that the old are more tired and less inclined to various spontaneous activities. They are also less mobile, and even if they are fit to travel around, they do so less frequently than people in the middle of their career. On the mental side, they feel more satiated with many types of life experiences and therefore tend to live a more quiet life. And, being less overstrained by family and job, they have more time and opportunity to turn good intentions into actual action, should they be so inclined. How about that for a sociological macro analysis?

SS: Yes, how would society manage without arm-chair sociologists? But if you are right, we have a quite different kind of causal argument here, pointing at factors that, more or less, apply to people just *because* they are old. They are therefore likely to be present in every generation when it reaches old age. These factors do not *follow* a cohort over time, but each cohort is likely to *enter* that stage in due time. Such factors are therefore in certain respects reliable instruments for prediction. All you have to care about is whether they are really causally effective for environment friendliness. If they are, you have identified a *life course effect*, as opposed to a cohort effect, and that should be better from your point of view.

ME: I am now beginning to see your strategy more clearly. You don't like the idea of results in the social sciences being mechanically applied, like results in the medical or engineering sciences, but would like me to take on some responsibility for interpreting the results before I use them for policy making. To make your point very clear, you present a result, or a somewhat regimented result, namely that older people are more environment friendly, such that it yields *one* prediction if interpreted in one way, namely as a cohort effect, and the *very opposite* prediction if interpreted as a life course effect. Right?

SS: Right.

ME: And now you want to raise your finger and exhort me: if you have a study categorizing people into different age groups and reporting different numerical values for the various groups, always look out for the difference between cohort and life course effects before you apply the result!

SS: Eh, yes. Do you mind?

ME: Yes, I do mind. What I need is not exhortations, but operative advice. How do I go about when I want to tell a cohort effect from a life course effect?

SS: Preferably, you should have a reasonably long time series of data. Were older people more environment friendly also 10, 20, 30 and 40 years ago? But usually no such thing is available. Then you will have to use your general background knowledge in an intelligent and systematic way. First you have to find out which individual properties there are that show up as environment friendliness and how causally efficient they are, and then you have to consider how these properties have come about: depending naturally on age, being imprinted at a early stage, being mere reflections of current trends, or something else.

ME: Which seems to indicate that there may be other possibilities than cohort and life course effects?

SS: Certainly. Migration, for example, tends to come in waves, and it leaves an impact on society's attitudes and values which is not limited to a certain age group, although it may affect different age groups differently. All this in addition to the values and attitudes of the migrants themselves.

ME: So, we have several kinds of effects, then, and they may all have something that speaks for them. How can I tell which is the true explanation of the attitudes of the elderly?

SS: Usually, it does not make sense to speak of *the* true explanation. The apparent social effect is, as a rule, the result of many different causal mechanisms, superimposed upon each other. And when there is occasionally only a single cause, you will have to use my old recipe to compare several studies made by different methods.

ME: And the other transition, from the now young to the future middle-aged? Same thing about that one, of course. Is the environmental concern of the young something imprinted on them and therefore something likely to last, or merely an expression of passing adolescent idealism? Hard for me to tell! And not really my business either! Things are getting complicated now, and I don't accept it any more that it should be my responsibility to make the synthesis. It would take much too long, and I suspect that there are many more pitfalls that I haven't heard of yet. You are either lazy or trying to avoid responsibility!

SS: I'll be quite happy to make the synthesis for you; of course I recognize that that is my job. But I won't be there every time you need advice, and often there will not be enough time to look deeper into things. The point is to gradually develop an instinctive feeling for these things. Isn't that the way you have learned politics? You must have met with unexpected or impenetrable political behaviour on many occasions in your early career, and only after express interpretation with some help from your political friends have you understood the real intentions behind. But now this process takes place automatically in your mind and you feel that you see through most smoke-screens without any conscious effort. The same thing is possible in science in a lot of cases. After you have done a few analyses you begin to recognize some general patterns and you gradually develop an intuitive feeling for what is superficial and what is more fundamental and you become able to smell a rat when there is one. That, I think, is as far as we should aim to get.

ME: And this is as far as I am prepared to go today. See you in a week's time.

Fourth session:
What variables should we look into instead?

ME: I had rather mixed feeling about last week's session. What struck me most immediately was the clarity and mercilessness with which you took apart my amateurish interpretations of age differences. Yet I felt that something was wrong without being able to tell exactly what. Only yesterday did I see it more clearly. We started out with what you called 'traditional sociological variables', and then we focused on one of these in particular, namely age. You then warned me to draw any conclusions from age differences alone and urged me to look at deeper causal relationship, if only for the sake of developing my instinctive feelings. But it now seems to me that the right conclusion would rather be to look for other types of studies, those that investigate less traditional variables, variables closer to the real causes. Aren't there any such studies?

And one thing more! You told me rather little about matters of substance, how things really are, and much more about formal matters, like what would follow from this if this were true. But I need plain facts also.

SS: Certainly there are many social scientists who look for 'less traditional variables' and many of them get quite interesting results. But you shouldn't speak condescendingly about the traditional variables just because of that. They form a sort of necessary basis. The real danger is to rest content with 'traditional' data without analysing the results any further. More complex variables are, however, difficult to compare from study to study, exactly because they are not standard. But they can also be very revealing if they are cleverly construed. Let me give you two or three examples, which, incidentally, will also supply you with some plain facts.

ME: Go ahead!

SS: First I would like to mention Anna-Lisa Lindén's study involving green identities. Identities in general are something one feels strongly about, something one creates, cultivates, and maintains consciously. They can be created or inspired by emotionally loaded information, by ethical, religious or nationalistic agitation, by commercial brand advertisements, and by striking models, and therefore tend to change in collective ways and are hence expected to be variable in time and between groups. If you want to create or change identities for the purpose of controlling green behaviour, it means that you have to choose different means for different groups and that you must be prepared to change strategy every now and then.

It is therefore very interesting and potentially helpful when Lindén identifies four types of identities, each with equal claim to greenness but rather different among themselves. She names them ecological identity, which is perceptual and small-scale, environmental identity, which is intellectual and large-scale, green economic identity, which emphasizes symbolic values and economic sacrifices, and green social identity, which is linked to social pressure and enhancement of self-image. The good thing about this is that the characterizations help to indicate what means one can use to reinforce these identities, the bad thing is that you can't find a single measure that will do the trick.

ME: Give me an example, then, of how you can reinforce then different kind of identities!

SS: Suppose you are dealing with a group of people engaged in the running of a wind mill for power production. To send out quarterly statements about how much carbon dioxide you have succeeded not to emit because of your commitment would reinforce an environmental green identity; to call a general meeting of all the people engaged in this particular mill would reinforce a social green identity, and so on. A variety of such strategies are in fact used by companies operating wind mills.

ME: OK, I see. Any more examples?

SS: Let's go on to look at Mona Mårtenssson's way of structuring her study. She begins by distinguishing five types of household cultures, borrowed from the so-called cultural theory. A household culture is a complex cluster of solidarity bonds, cultural ideas, and behavioural strategies. The five types are called fatalistic, hierarchical, egalitarian, individualistic, and autonomous household cultures, each defined by characteristic features, just like Lindén's identities. But unlike these they are more unconsciously received, more conventional, a sort of habitual arrangement of a major part of every-day life. Perhaps they are therefore less susceptible to deliberate manipulation, more reflecting the spirit of a period or of a social context and hence changing more slowly.

But household cultures form only one of three life contexts. The others are life course phase and social gender order. Common to all of these are that they are not individual properties, but reflecting the place of an individual in a social or societal context. They are therefore liable to change more in response to structural changes in society rather than to individualistic persuasion measures. It turns out that e.g. purchase of eco-labeled products is better explained by household culture than by life stage or gender order, while car driving shows the opposite pattern, being quite unrelated to household culture and also to environmental attitudes in general.

ME: So a new concept, or a new variable, may explain certain things well but others not at all. Can you still say that you should in general try to go for 'non-traditional' variables?

SS: Not exclusively, as I just said. In fact, it is a rather complicated process before you can use innovative research for policy making. First there has to develop an agreement between many scientists about which variables are important, interesting or potentially useful, so that there will be enough studies for a more complete picture to emerge. Then the new variable has to be placed in proper causal chains both forwards and backwards, meaning, on the one hand, that you should be able to tell which environmentally relevant behaviours it does affect, and, on the other hand, how it can be controlled, influenced, or manipulated. Failure on any of these counts will make the research uninteresting from the point of view of applications, no matter how much internal interest and explanatory power it may possess.

ME: That's a fairly bleak picture you paint for unconventional and innovative social research! Was that really your intention – you sounded rather enthusiastic before.

SS: Not altogether – perhaps I got carried away again. As a social scientist myself, I perhaps tend to look more for fundamental explanation than for immediate application. But I find consolation in the fact that it is explanatory powerful theories and results that almost always pay off the most in the long run. As I have said irritatingly often, new results can only be interpreted against a background of a large number of related studies, and explanatory successful theories provide more productive contexts, whether or not they are applicable on their own.

ME: It seems to me that you are advocating something like this: we need to accumulate a kind of research infrastructure, a collection of related studies, preferably innovative and explanatory powerful, without looking too much at immediate applicability. It shouldn't be too narrowly focused, but certainly not aimlessly scattered either. Once we have that background structure we may well commission more pointed studies, more directly addressing specific application problems, with reasonably good hope of interpreting them correctly enough for direct applicability.

SS: I agree. The great advantage with studies having a theoretical foundation is that they easily form links to other studies – the raison d'être of theories is after all their generality – and therefore more efficiently contribute to the web-like infrastructure you mentioned. Their disadvantage is that they seldom lend themselves to immediate applications. So we need studies about particulars as well as generalities, but in the right order.

ME: OK, let that be the concluding note for today. Next week I would like to know, not only why we should study more fundamental variables, but which ones!

A week later:
What about information?

ME: I have been reading some of my notes from previous sessions, and I am struck by the fact that you have talked so little about information. I suggested the topic once, but you ducked and said that it was a tricky business and deserved a treatment of its own later. Maybe the right time is now?

SS: Yes, why not. But information has so many aspects. Why did you come to think of it?

ME: Well, first of all information seems to be very important. We launch information campaigns about all sorts of societal matters when we think that citizens should be enlightened, and we think that a free press is a cornerstone of democracy. Secondly, there must be lots of research about information already, considering the amount of money that is spent on advertising and marketing. Some of it may be very specifically tuned to that purpose, but certainly there must be something to learn from it for the environment also.

SS: There has certainly been a lot of studies about information, but perhaps of a different kind than you think. So we must begin by making distinctions. First of all you must ask *why* information is important. There are two common answers to that question. Anyone uncontaminated by social science would certainly say that information is important because it produces knowledge in those who receive it and knowledge is valuable. But if you look at the majority of studies of information, you will find that they almost exclusively try to measure the change in *attitudes* that information produces, even if they may admit that this change is achieved *via* a change in knowledge. Which is most important for us to study, knowledge or attitudes?

ME: Neither, I would say! What we are ultimately interested in is behaviour and how information affects behaviour. Then both knowledge and attitudes become mere transit concepts, intermediate stations linking information with behaviour but not of real independent interest.

SS: Wrong! Remember the note you made a couple of weeks ago, not to rest content with describing the immediately observable but to discuss underlying causes? If you just focus on behaviour you will learn much about the cases you actually study but will have little to say about the range of applicability, and you won't have the guide to policy-making that you want. Both knowledge and attitudes are more fundamental than behaviour and each deserves to be studied in detail, although knowledge, being purely cognitive, is more immediately connected with information, while attitudes mix cognitive with evaluative elements. So I suggest we come back to attitudes later. The point I really want to make is the distinction between information and knowledge.

ME: Aren't they more or less the same, knowledge being just an honorary label awarded to that which is more worked out, better organized, and perhaps also a little classier?

SS: I understand that you can see it like that, but there is an important distinction: information comes in separate chunks while knowledge is holistic. Social scientists tend to look at information as an abstract counterpart of a daily newspaper – it comes from a particular source, is broadcast over the country, some of it falls on barren soil but some is duly received and then adds a piece to the jigsaw puzzle formed by that person's knowledge. When thought of like that, it is easy to regard information as a commodity and to focus only on its internal properties.

But knowledge is a different matter. It is a state of mind in the receiver, and a given piece of information may trigger quite different reactions in different receivers. It may be strange and incomprehensible to some and stay as an isolated item in his or her mind, or, more likely, soon be forgotten, or it may turn out to be the missing link that suddenly reveals a new solution to an itching problem.

ME: Which makes it much harder to use information as a means to influence people and to calculate its effects!

SS: Yes, and it also makes it much harder to do research about knowledge! That's why there is so little research about knowledge proper, as opposed to information and attitudes. You must find a way to represent the *content* of knowledge, which is not a simple numerical measure, and also how a content interacts with an existing state of knowledge. Such representations do exist in logic and the philosophy of science, but seem to be unknown in the social sciences, and consequently there are no empirical studies around.

ME: Nothing for the near future, obviously. What, then, is the second best? Perhaps to study attitudes or information as a commodity after all?

SS: Perhaps. As for attitudes, there are many studies about them, and they show that it is certainly possible to influence attitudes with information, particularly if the information has an evaluative component, even if only implicitly. But it is doubtful whether such effects become very entrenched in the mind of the receiver and hence whether they last very long.

As for information considered separately, you must make a distinction between information about general conditions and mechanisms and information about single facts. To understand properly information about how heavy metals, like mercury or cadmium, affect the human body requires some amount of chemical and biological education, but if that is missing you can still form some kind of simplified picture in your mind if you are motivated. Information of this kind, even if launched through an intense campaign, will reach only few, but once received it will have predictable effects and will form a rather robust structure in the mind which will govern the interpretation of future data and direct both individual and collective action in certain directions.

In contrast, the information that this kind of battery contains cadmium will evoke quite different reactions in different people. To some, it will be simply redundant. Many others will discard the information as irrelevant, uninteresting or incomprehensible. Others still will feel uneasy because they have learned that cadmium is a bad word in environmental circles. It all depends on what soil this seed will fall. Many will no doubt act 'correctly' and dispose of the battery in a special way (if it can be done conveniently) but the commitment to such behaviour is probably not very strong and it may well be overturned if some contradictory evidence shows up. Compare for example to alcohol: it is bad for many things but good for some, and there is a continuous flow of newspaper reports about its effects. What information will be noticed and remembered by an average individual depends much on prejudices, selective perception and wishful thinking, but for someone well acquainted with bio-chemistry some (and different) reports will stand out as interesting and relevant and worth acting on.

ME: Less academically: information about how things hang together is valuable but hard to sell; information saying merely that this or that is bad for the environment may well have an immediate effect but is superficial and likely to be superseded by new information. Right?

SS: That's my opinion, yes, but only the last part is sustained by empirical studies; the first part is eminently reasonable but hard to do research on.

ME: What sort of empirical studies sustain the last part?

SS: Studies relating competence in a field to the ease with which one is influenced, for example. They show that the more competent you are, the less likely it is that you conform to standard packages of views, like the official views of political parties, and the harder it is to influence you, at least by simple messages. It has for example been repeatedly shown that American election campaigns succeed in changing attitudes mainly among the less educated. Interestingly enough, however, there are some signs that this is not always true in Sweden, possibly indicating that Swedish campaigns focus on more complicated messages.

ME: Why is that? Do you mean that the less educated are more gullible? In my experience they are not! They have firm convictions and are not afraid of showing them. You can even say that they are sometimes downright stubborn!

SS: About things with which they are accustomed, yes. More abstractly speaking, there is a well-known theory from the late 1950s by Leon Festinger which has been widely accepted, at least in its revised version from the 1960s. It says that the governing principle for how people notice and pick up information, interpret it, and retain it in memory is that they tend to avoid cognitive dissonance, that is they try to avoid for example contradictory beliefs or unresolved tensions in their mind as far as possible. This explains both the stubbornness about interpretations of the well-known, because opinions that have settled in a cognitive equilibrium can't be changed one at a time but have to be restructured more thoroughly and the less educated are not imaginative enough to see how to do that, as well as a certain gullibility about the unknown, because there is little pre-existing knowledge for the new information to connect on to, so whatever seems reasonable in the absence of alternatives goes down.

ME: So if you wish people to believe something contrary to habit or popular wisdom or otherwise unexpected, it is no good to just say so. You have to do some destructive work first, question or break up habits of thought or conceptual schemes in order to create an openness to accept new information. Isn't that exactly what Anders Biel says in his article? You have to upset people's cognitive equilibria before they are ready to take in new information!

SS: And the question then becomes of course how to do just that. That, again, has to do with the kind of information involved. There is another psychological theory, developed by John Cacioppo and Richard Petty, which places emphasis on mental elaboration. The more a piece of information is elaborated, the more it becomes entrenched in the mind and the more likely is it to last and to govern the interpretation of further information. The theory focuses on the likelihood of elaboration, which is of course higher if the subject is interested, informed or well educated. It also distinguishes between two types of information, namely 'cues' and 'arguments'. A 'cue' is a piece of information that carries more associative value than actual content – like 'Mike is careless with the truth', 'we will solve all your banking problems' or 'car driving is bad for the environment' – while an 'argument' is a more complex informational structure, with a well-defined content and a definite point.

Cacioppo and Petty find that cues have a greater impact on attitudes if the subject is less interested, less educated or otherwise unlikely to elaborate the information, while their effect is much smaller or even opposite for subjects likely to elaborate. Arguments show the opposite pattern, and more so the stronger the argument is. This is one of the few successful attempts to categorize information according to its content, although it does so only by classifying along the scale weak/strong or simple/complex and not by actually representing the content. What it can explain is thus only *that* someone will be affected by the information, not *how*.

ME: So you have to send different messages to different target groups. But you can't prevent the interested and educated from listening to the message intended for the uninformed and uneducated, and if the effect may even be reversed, how do you manage that?

SS: Select your channels carefully and adopt the message to the channel. Unfortunately, this strategy has many drawbacks – it contributes to stratification in society and tends to marginalize popular media from serious decision making, even if these may exert considerable short-term opinion pressure in individual cases.

ME: So, what you suggest is that there are two reasons why information about more fundamental environmental, economic and social mechanisms is more important than slogan-like exhortations what you should do in order to be nice to the environment: it has lasting and more profound effects, and it affects more important groups of people? But is there really any evidence for that?

SS: Not very direct evidence, because these things are difficult to study, but quite an amount of indirect indications. But many of them become visible only when you look at long term trends.

ME: Let's do that, then!

SS: I have to start with an elaboration of some examples.

ME: Then we will postpone it to next week.

How to interpret cross-sectional and longitudinal data

ME: Why do you scientists always use so many esoteric metaphors? You don't mention time at all in the note you gave me about today's topic, and yet I suppose that time is what it is all about. I take it that cross-sectional for you means something like taking a flash photograph of society at a given point of time and then looking for interrelations between the kind of things you can see there, while longitudinal means following something through time. Am I right?

SS: Yes, of course. Isn't that obvious and quite literal?

ME: No, it is not. I don't mean to quibble about words, but I see a risk in social scientists trying to appropriate the very useful distinction between cross-section and longitude for limited use in relation to the time dimension only. But I can think of many other good uses. I could for example study the pollution in a river. The state of the river – flow, speed, prevalence of fish, oxygen content, presence of various pollutants and the like – at any given point could be studied I relation to the surroundings – topography, inhabitation, crops grown, industries – which would be cross-sectional, or you could study how these characteristics change when you follow the river from its source to its mouth, which would be longitudinal. And the two kinds of studies will reveal different kinds of facts about the river. Or one should perhaps rather say that one's prejudices or expectations about what factors are important will determine, often unconsciously, which kind of study comes to mind. Is it a similar thing you want to tell me about social time series data?

SS: Well, yes. Only I am a bit confused where to start since it seems that you've got the point already.

ME: No, I have spoken in general terms only, saying *that* the two approaches differ. You can fill me in on *how* they differ when it comes to social data. I mean, which factors reveal themselves only in cross-sectional studies and which ones in longitudinal studies?

SS: I can't. It's a different logic: you can study some factors – for instance sex, age, and all those 'traditional' variables we talked about before – in cross-sectional studies and find that several of them have some explanatory force, and yet they are clearly insufficient to explain variations over time. There has to be something else as well. So I can tell you what does *not* reveal itself in longitudinal data, but not what *does*. We have to look for a residue, but need not do so blindly, for we know what temporal pattern it has to follow.

ME: Even more general terms! Matters of fact, please!

SS: All right, I will borrow some examples from *On the Mechanism of Opinions* by Torsten Österman, who has, I think, treated these matters very nicely. He has been mainly concerned with political opinions, and I will use some figures about Swedes' attitudes towards the European Union. Sweden joined the EU after a referendum in the end of 1994, but a possible membership had been hotly debated many years before that. People were asked whether they considered it 'absolutely required' that Sweden joined the EU and the answers were graded on a 4-point scale. The interviews took place from 1988 to 1994, more frequently in the last year.

Cross-sectional data were very consistent. On *all* occasions men were found to be more positive towards the EU than women, older (65–85) more positive than younger (0–35), inhabitants in urban areas more positive than in rural areas, and the academic classes more positive than the working classes. However, the mean response varied substantially over the years, and it varied uniformly for all the subgroups mentioned. In fact, at *any* given point in time, the difference between the most negative and the most positive group was considerably smaller than the difference between the maximum and minimum over time for any given group. Over time, the opinions of the various groups went up and down keeping very much in step, but also always keeping their relative positions.

ME: So obviously some time-dependent factor had a greater influence than membership in any of the groups you mentioned. Any idea what it could have been?

SS: Idea yes, conviction no. It seems to me that different aspects of the EU came into focus at different times. First, there was much talk about the EU as a peace project, which generated quite favourable attitudes. Then economic co-operation came into focus with some anxiety for domestic industry but also hopes of lower prices. And late in the day it was the threat to national sovereignty that was on the carpet, generating negative attitudes. On all three counts I think that people were reasonably well informed. Yet, if I am right, the varying factor was information of a kind, but not content accumulating over time, but a flow of knowledge of details, conceivably deliberately managed by political actors, which added little content but served to focus interest and public discussion on certain dimensions of the EU question, which was in itself so complex that few could form a balanced picture of it in their minds.

ME: I see. Not lack of information, but a sort of mind-framing that forces selective recall of what you already know.

SS: And, not to forget, selective reception of new information!

ME: Complexity as such seems to be a problem, then, at least when it comes to popular vote. Of course, you admitted that you were only venturing an opinion, but the mechanism you suggested is certainly something we have to look out for, whether or not you are right in this particular case. Perhaps this is the right time to remind you of one of your hobby-horses: to look at several studies and compare them in order to get an idea of the range of applicability of a line of thought. Any other results of the same kind?

SS: Yes, perhaps. I am thinking of another of Österman's examples, this time about studies of subjectively perceived quality of life. Such studies have been common since the early 1970s. If you look at cross-sectional data, the most important determinants for a perceived high quality of life are what can be regarded as components of this concept, that is 'soft' factors like good relationship to friends and family, quality of free time, work satisfaction, and the like, whereas 'hard' factors, like (subjectively perceived) personal economy, are poor predictors of life quality. But if you look at variation over time you will find instead that satisfaction with life in general correlates very strongly with perceived personal economy while e.g. quality of family life or work satisfaction explains next to nothing.

ME: How do you explain that?

SS: Too complicated, and too speculative for our discussion today. But one could make a parallel to the way many studies of environmental behaviour are designed: there is a host of cross-sectional data about the relation between environment friendly behaviour and various personal variables, both traditional and less traditional, but there is much less about relations to more general variables. My personal favourites are three: ease of handling, economy, and trust in the meaningfulness of an action. The last one, I think, is strongly linked to information in the deeper and more complex sense – to 'arguments' in Cacioppo's and Petty's sense – and the two first ones are more structural, depending on society's organization rather than on personal traits. I would like to see many more studies of how such factors relate to environment friendly behaviour. Such studies could be time series studies, but could also be longitudinal in your more general sense. That would perhaps even be preferable because results would come sooner.

ME: These examples were very interesting; it seems to me that we are actually moving ahead now. You remember that I began our relationship by complaining about social science results being so context dependent and of so limited applicability. I meant that in relation to the substantial changes that nevertheless take place in society and which are only very partially explained by the social sciences. Perhaps we have found at least the direction where we should search to remedy this. Can we sum it up as a memo for future research plans?

SS: I can think of no better way than to quote at some length from Österman's own conclusions. He says:

> The measurement of an attitude is a measurement of a reaction, often basically affective in its nature. The reaction is typically a summary of an impressive amount of data and experience, but the measurement value does not allow us to see much more than the single (affective) summary value.

In any given moment, it will be that summary that is most visible, and in the absence – or the relative absence – of variations other than those between objects, the attitude will appear more or less as a trait. The covariation of that trait with other traits such as class, gender, life-style, age cohort and the like, will seem to indicate causal links to the attitude. The reasons for that view are not invalid, but dependent on what its purpose is. If it is to make short term predictions about election turnouts, what people might think of an environmental issue or how they will react to busing or foreigners, the reasons may be sufficiently wise. If, however, the purpose is to understand long term change, how an attitude is constructed, or the processes and mechanisms behind those changes and that construction, the use of background data for its short-term predictive value is not wise at all, and in fact can be quite misleading. Those mechanisms become more visible in empirical data as the time perspective and number of content-differentiating indicators grow.

I think that Österman here touches upon an important problem which has been systematically tackled in the natural sciences but is largely neglected in the social ones, although it has great methodological significance. If you permit I will first make an analogy with physics?

ME: All right, go on!

SS: Objects near the surface of the earth are influenced by a great number of forces, causing them to move in all sorts of ways. No matter how diligently you observe and record the movements of falling leaves from trees, grains of sand on the beach, and water drops in a fountain, you will only see the gross effect of all the forces and you will have a very hard time finding out even the most basic underlying mechanism, namely gravity, from merely statistical analyses. The natural sciences have designed a device, called *experiment*, which keeps all factors constant except one in order to study the effect of that factor alone. A birch leaf that falls in vacuum is exposed to gravity only, but in real life any gust of wind is much more important than gravity for the movement of the leaf. If interviewed, each falling leaf would say that the wind is the most important factor governing its life. But winds do cancel out in a way that gravity doesn't, and if you look at the ground after a while you will find that it is gravity that is all-important in the long run.

The same holds even more for the social sciences. What you can observe is merely the gross effect of innumerable factors and it is next to impossible to identify independent underlying mechanisms by statistical analysis alone. And the social sciences can't remedy this by straight-forward experimentation to nearly the same degree as the natural sciences because it is often impossible and sometimes unethical to manipulate people with sufficient precision and in enough numbers. The social sciences have to seek compensation in more sophisticated study designs and concept formations. But it should be remembered that manipulation is not the goal in itself – the goal is to study the variation of one variable when all the others are kept constant, and should it so happen that data reflecting that do exist naturally somewhere, an efficient process for identifying them would be even better than an experiment that duplicates them. The division into cross-sectional and longitudinal data can be seen as an approximation of this – the variables are divided into two groups, time-dependent and not time-dependent. Cross-sectional studies keep the time-dependent variables constant and study the relations between the others, while longitudinal studies do it the other way round.

ME: That was quite a lecture! I'm exhausted. See you in a week's time.

Can we assimilate all the information we get?

ME: You spoke about your three favourite candidates for factors of long term importance: ease of use, economy, and trust in the meaningfulness of what you do. How does that square with your insistence that fundamental information about how things hang together is also important in the long run? It relates only to the third of your factors and not to the first two.

SS: The first two enter the picture at a later stage. Fundamental information is not only used by individuals when they perform their individual actions. It is also, and perhaps mainly, used for designing routines, creating organizational structures, and forming basic values. This is, by the way, why such information exerts influence only in a rather long time perspective. The point, therefore, is not to look too much at immediate effects of small pieces of information, but at accumulated effects of successive increments of information.

ME: Is there a difference, really? Isn't the accumulated effect just the sum of the many small immediate effects?

SS: A ten mile walk is the sum of a great number of individual strides, but not any sum of so many strides make up a ten mile walk. They also have to be more or less in the same direction. When each step is taken randomly you may well end up near where you started. This is one of the reasons why fundamental, theoretical and complex information has an enduring importance. Once received, digested and accepted, it serves as an interpretation frame for new information, particularly about more limited facts, and thus identifies a direction. You have probably heard about paradigms as mental structures that govern your whole world view. This is the same thing on a smaller scale. But fundamental knowledge is not the only thing that governs perception – ideologies, both political or scientific, do it too, and probably more efficiently.

ME: So why bother about information? Just see to it that you favour the right ideologies instead!

SS: Ideologies are already there and it is not clear that anyone of them is 'right', and of course they can't be judged by environmental criteria only. What you need to study is how ideologies 'transform' fundamental knowledge. They usually don't deny facts but force certain interpretations on them in a way that frames attention and interpretations of future facts and may therefore point out a direction which profoundly affects the accumulated effect of knowledge.

ME: So you mean that ideologies are downright harmful? Isn't there any way to bypass them? Why can't we just face the facts, construe a set of rational solutions or action strategies, and then maybe ideologies may enter the picture by providing the values by which we choose among the options?

SS: No, ideologies are not downright harmful. They are not downright useful either, but they are but one example of a common and necessary phenomenon: the need to reduce a complex reality to manageable proportions. The human mind can store immense amounts if information in memory, but the capacity of our 'working memory', the focus of consciousness where actual intellectual work takes place, is severely limited. The psychologist G.A. Miller once wrote a famous article called 'The magical number seven, plus or minus two'. He argued, and some say proved, that people can hold at most 5–9 'chunks' of information in the active part of mind simultaneously, and even that requires intense concentration. Doctors who routinely take a lot of tests nevertheless in practice use only three or four criteria when making diagnoses, and you find similar figures for other professionals. It seems that what goes under the name of thinking comes in two varieties: passively allowing a chain of associations to form itself, which could successively engage any number of facts, or actively working at an integrated understanding of quite a small number of facts.

If you allow me a little digression, I would like to quote Bertrand Russell here. He holds that the first kind of thinking is largely done in language, where words evoke associations to other words to form a discourse (although he didn't call it that), which is, of course, incapable of transcending the realm of thoughts already encoded in language, whereas the second kind of thinking is about things in themselves, independently of any linguistic representation, a kind of structural beholding. He says about philosophical logic: 'the subject matter [...] is so exceedingly difficult and elusive that any person who has ever tried to think about it knows you do not think about it except perhaps once in six months for half a minute. The rest of the time you think about the symbols, because they are tangible, but the thing you are supposed to be thinking about is fearfully difficult and one does not often manage to think about it.' The importance of the distinction lies, of course, in the fact that only the second kind of thinking results in such fundamental knowledge as will form enduring interpretation frames of the kind we have been talking about.

ME: Very amusing! But not quite in line with your own thought, is it? Because you seem to have a vision of giving 'fundamental, theoretical and complex information' to the people, haven't you?

SS: I realize that that is possible only in a sense, and you will see how. It is easy to interpret Miller's result in a purely negative way, as revealing a limitation to the mind's capacities which in turn results in oversimplification or superficiality. Ideologies are sometimes mentioned as examples of that, being regarded as necessary but deplorable. But it need not be so. The best way to turn Miller's result into something positive is to try to form as big 'chunks' as possible.

ME: Why do you use such sloppy language, I mean 'chunk'?

SS: It's not sloppiness. There is no established word for what I mean. A 'chunk' is a strongly coherent collection of facts which can be apprehended as a unit. As long as it can be apprehended as a unit it only occupies one slot in working memory. 'Bits' or 'pieces' would suggest fragments carrying a minimal amount of information. That a patient has rapidly rising fever, headache, nasal catarrh, high sedimentation rate, running eyes and white spots inside the cheeks are separate bits of information, but that he has got the measles is a 'chunk'. The point is that a chunk may carry quite an impressive amount of information if it is well integrated and conceptually established. Structuring facts into larger meaningful units is a well-known technique for improving one's memory.

ME: Now I see the pattern! You will argue once again that 'fundamental, theoretical and complex information' is important because it makes you understand how things hang together and can therefore serve as the backbone of a really big chunk if you work it up with suitable details, and in that way you can transcend the magical number seven.

SS: Yes. But it is hard work. It takes concentration and motivation, as Russell would have agreed. The main difference between those with high and those with low mental capacity lies in their ability to form integrated chunks rather than in their access to many facts. That's a hint you may pass on for free to your peer in the Ministry of Education!

ME: No, I won't. It would hurt the self-image of her crew as fore-runners in the information age. But haven't we drifted away from the practical consequences?

SS: We'll soon be back on track! Let us assume, with Russell, that most people will work with quite small and unintegrated chunks most of the time. This means that we have to imagine a person with quite an amount of relevant (and irrelevant) information stored away in memory somewhere, of which, however, very little is present in the focus of consciousness at any one time. The process of thinking, as reflected also in discussions and presentations, can be thought of as the continuous inflow, processing and outflow of information in active working memory. Outflow is easy – it's just what's pressed out by new inflow and limited capacity. And actual processing is very taxing, so actually it is the inflow that determines the train of thought in practice. So our main problem is to explain what governs the inflow.

First of all, thought is, to a substantial part, a self-sustaining process. Elements in mental focus have associative links to other elements, which will succeed them in focus. Brand advertising and propaganda therefore aim at establishing and reinforcing such links, often by connecting to emotionally loaded ideas: 'Guinness is good for you' or 'Would you buy a used car from this man?'. In their simplest forms such techniques just stick an emotional label, such as security, dependability, danger or anxiety, on some phenomenon. But there are also more sophisticated cases, many of which involve non-trivial amounts of factual knowledge: nuclear power evokes anxiety by associations to, say, radiation, cancer, genetic mutations, deformed children, and irresponsibility to future generations, waste incineration is linked to toxic and cancer-causing chemicals such as dioxins and polyaromatic hydrocarbons, while wind power suggests clean air, renewability, idealistic enthusiasm, and the like.

ME: Is all this bad? The first kind is a bit trivial, of course, but isn't the second kind all right? It's by and large correct, isn't it?

SS: It need not be good just because it is correct. I am trying to describe the most common kind of thinking, and I see certain dangers with it. It is non-constructive, it causes positions to become entrenched. Arguments tend to be repeated over and over again with little chance of leading to something new. And it makes a large number of people think in parallel tracks, barring that variation which is necessary for evolution.

ME: And what can you do about that?

SS: Make people think in larger, more integrated units. That will make them more autonomous. Emotional labels won't stick as easily, because there will be several links leading to the same unit, carrying different associations and therefore creating a kind of ambiguity which is likely to be resisted and which may even trigger a reinterpretation of the content. And the number of links leading out from the unit will also be larger, making the thought process increasingly ramified and unpredictable.

ME: But you're talking only about self-sustaining thinking, which is a very special case. What about when something extraordinary happens, like the Chernobyl accident, which forces itself on everybody's thought?

SS: Things happen to us all the time, even if you only sit at a café looking at people passing by. Dramatic or not, an outside event brings elements buried in memory, sometimes deeply, to the fore of mind, and then the same associative process starts again, but from a different starting point. The real difference is not between self-sustaining and externally induced thought, but between cognitive in-depth processing and thought guided by trains of associations.

ME: It seems that Miller's greatest insight was the introduction of the concept of chunk rather than the limitation of the mind's capacity as such.

SS: It's the limitation result that bestows importance upon chunkiness. But the limitation has other consequences too. It means, for example, that it is very difficult to be entirely consistent in one's values and judgements. Within a well integrated chunk, there is hardly room for obvious inconsistencies, but local consistency in each chunk is no guarantee for global consistency. We start to build up our understanding of different phenomena in independent ways, and we form each sphere of knowledge into a coherent unit in its own right with its own central concepts and established truths. Clashes are sometimes not even noticed because they are couched in different words, and even when noticed they are not really worked at because that would require too much reorganization of the established units.

ME: Is that a general impression of yours, or can such things really be scientifically established?

SS: There are two kinds of studies which are particularly relevant, in complementary ways. One flourished in the philosophy of science a couple of decades ago and succeeded, mainly by sweeping historical examples, in showing that even within science proper are anomalies accepted for quite some time before such changes occur as require the breaking up and reorganizing of established coherent theories. I am referring, of course, primarily to Thomas Kuhn and his theory of paradigms and scientific revolutions, but also to Michel Foucault and his ideas about different *epistemes* succeeding one another in the history of human thought.

The other kind of studies is much more rigorous and is known as decision theory. Although it began as a prescriptive theory about rational behaviour, its theoretical results have found many uses and they are certainly of relevance here. To be fully rational means, among other things, to be globally consistent, both in one's values and one's beliefs. Decision theory then proceeds to define both necessary and sufficient conditions for global rationality, given certain values and beliefs, but without taking any stand whatsoever as to what these values and beliefs should be. You could say that it concerns itself exclusively with process rationality and not at all with what is sometimes called value rationality.

You can then compare the prescriptions of the theory with how people actually choose. Not surprisingly, there are clear and systematic violations. You can interpret this in two ways. Either the theory's conditions are too strict and do not really reflect true rationality, or people act in fact in inconsistent ways. There has been much discussion about the first alternative, but, sad to say, most of it has been based on misunderstandings. People have misinterpreted the internal jargon of economists and read too much into it.

ME: Well, explain, then. Or do you just want to impress the ignorants?

SS: It will take a little patience from your side, but I am sure you can produce that. Let me first state the pivotal theorem without any technical jargon as best I can. It proceeds from a couple of innocent (but essentially global) minimal conditions on rationality. They are things like if you prefer x to y and y to z, then you prefer x to z, that equally preferred things can be substituted for each other, that your preference for a lottery where you win either x or y lies between your preferences for x and y, and so on. The conditions *prescribe* nothing about your preferences; they only say that *if* you have such and such a preference, *then* you should have so and so a preference in the name of consistency.

ME: Clear enough! But how can anything interesting come out of that?

SS: Perhaps nothing interesting at all. What comes out – and it comes out very consistently and systematically – is the empirical finding that people's preferences do not obey even such simple rules. It has nothing to do with the content of the preferences, but only with their internal consistency. It proves that global consistency is very hard to come by, quite in line with what we have said about chunks. This is not an insignificant result, because inconsistencies make people susceptible to wishful thinking and to being manoeuvred and they appear systematically in interesting ways, but the result is in no way revolutionary.

ME: That's not the way I've heard decision theory described. I've heard a lot about maximizing expected utility, and that seems pretty prescriptive to me. And also narrow-minded!

SS: 'Maximizing expected utility' is exactly the jargon I was referring to, and that's where all the misunderstandings are.

ME: Then you had better explain, hadn't you?

SS: OK, but you will have to put up with some of the technicalities then. Decision theory is very easy to test, because you can derive a very powerful technical theorem: if a person's preferences satisfy the conditions, then there can be shown to exist two numerical functions with very seductive properties. One function has all the formal properties of a probability function and the other assigns values to outcomes, high values to preferred outcomes and low to less preferred, on the whole. The theorem says that if you order all options in preference order, then you will end up with exactly the same ordering as if you ordered them according to the expected value of the second function with the first function used as a probability function. If you leave those functions totally uninterpreted, this only means that you can find indicator values and stick them onto the options in a consistent way.

The jargon of the trade is to call the first function the person's subjective probability function and the second his or hers utility function. In that jargon, it is true to say that to be rational is the same thing as to maximize one's expected utility.

But that jargon is known also to those outside the trade. They find it very hard to think of the functions as uninterpreted indicators. It is much more tempting to think of the first function as the person's true subjective probability function and of the second function as a measure of the person's perceived utilities. Perhaps the person was not aware of these function, and the great achievement of the theory is that it has somehow revealed them. To say that a rational individual always chooses so as to maximize his or hers expected utility then takes on a new meaning. To many, such a statement is blatantly and revoltingly wrong, and the theory is repudiated as egoistic, materialistic, economistic, mechanistic, or what not.

ME: I agree! They maximize something, and even if what they maximize isn't exactly their utility, it must be pretty close. But certainly many people are driven by quite different forces. They often act for very idealistic reasons, in particular when it is about the environment.

SS: No, you are wrong, and for two reasons. The first one is rather obvious, once you come to think of it. Even if the function would measure a true utility, the theory is completely indifferent as to *whose* utility it is. A consistent environmentalist, who always acts for the best of the environment regardless of any individual interests, will act in perfect accord with the theory. And so will he who acts in the interest of the nation, or for the sake of the poor, or to do God's will, as much as he who acts from pure egoism or harsh materialism. All the theory cares about is the consistency of the preferences, not whose interest they reflect, nor, indeed, whether they reflect any interest at all.

ME: Good point! I was wrong about that. Or perhaps not I, but some sort of diffuse opinion, which I uncritically took over. Sorry to be gullible before prejudiced opinions, but that's a professional disease in a position where you very seldom have time to go into any depth. Or perhaps there is something to the criticism after all? To maximize something, even if it is a noble goal, is a very mechanical way to go about things. Certainly we are not as calculating as that?

SS: The central theorem says nothing about calculations, and the person need not have any conscious notion of maximization at all. That was in fact my second reason. Recall that when I explained the content of the theorem, I took not only great care to call the functions just 'functions' without labels like 'probability' or 'utility'; I also cautiously said that the person's preferences were *as if* he or she maximized a certain value. The theorem *starts out* with preferences, which can be based on just anything – aroused emotions (noble or disgraceful), prejudices, or cold calculations. It *then* states that those preferences are *as if* the person maximized a certain expected function value. But the functions, and the formula, are pure constructs by the theory, revealed *from* the preferences and not providing reasons *for* them. There is no reason to assume that the functions or the formula correspond to anything real inside the mind of the person. This is also why any inconsistency revealed by the theory can be regarded as much an inconsistency in values as one in facts.

ME: You said that this theorem made it very easy to test decision theory. How?

SS: You simply ask test persons what preferences they have among a selected number of options. Assuming the consistency conditions, the theorem makes it easy to find out what preferences they should have among a second set of options. Then you ask them directly about this second set, and very often you find that they answer something different. Conclusion: quite frequently the consistency conditions are not met in real life.

ME: Or they just didn't understand the consequences of their own standpoints. That happens, you know, also in my business.

SS: Yes, of course. But that's just a special case of inconsistency. You may attach great value to certain general principles, say about democracy or freedom or perhaps about something less grand, and yet dissociate yourself from some of their specific consequences, precisely because you don't see that they *are* consequences. If you are made aware of this, you may react in either of two ways: change your attitude to the specific consequences and come to regard them as minor drawbacks of a greater good, or to reject the principles because they weren't as evident as you first thought. But in either case what you do is to straighten out an inconsistency. And this happens all the time – the more so the more reflective you are.

ME: If this happens all the time, why is it so? And why are some people so dead certain about thing?

SS: This is not for decision theory to say. But I think that we are back where we started today – it's all a consequence of the fact that we learn about things in local chunks; global inconsistencies have to be taken care of later, but first they have to be noticed, and the big difference between people is between those who notice and those who don't.

ME: That's a disappointing result, making it pointless to seek more information and so rather useless for my purposes. Is that how I should interpret what you say?

SS: No, not altogether. As I said, the violations are systematic. There are two types of situations where they typically appear: when very small probabilities are involved and when the decision is made in two or more steps where the options are structured differently. The first case covers both lotteries and risks for catastrophic events, and in both cases the probabilities tend to be overestimated, but of course it is the latter case that is most important in our context. It is hard, mentally, to envisage very small probabilities in a numerically exact way unless you work with them very closely, and therefore their chunks tend to be small and dominated by the character of the event rather than its probability. It is easier to integrate the pleasures of a million-pound prize or the horrors of a nuclear accident with the conditions of your future life than to integrate very small probabilities with probabilities you experience in daily life.

ME: But all this is certainly well known to those who run lotteries or some insurance business, who profit from it!

SS: And to many environmentalists! But you're right, these particular results may not be very exciting in their own right. What is perhaps more important is the second case, the many-step decisions. I am not thinking of trivial cases, like when you take part in a competition where the prizes are lottery tickets, but of cases where different mental strategies apply in the two (or more) steps. If, for example, you consider an operation which almost certainly will relieve your chronic pain but yet involves a small probability that things will get even worse, then security considerations will be important, and you will spare few efforts to eliminate or at least reduce the little risk there is. If, however, the same situation is imbedded in a broader context, for example if the operation will only be feasible if some blood tests come out positive, which is *a priori* rather unlikely, then the whole situation is rather marked by resignation, although with some hope of success. In such a case, the probability of success becomes a more important factor in the decision, psychologically speaking. It is easily shown that this leads to inconsistencies.

ME: So what? Why shouldn't people be allowed inconsistencies if it makes them happier or at least feel all right?

SS: On the small scale, limited to the personal sphere, inconsistencies may perhaps be harmless. But they also make people helpless and at the mercy of other people's ways of presenting things. Several of the most famous results in decision theory show just that: you can control people's preferences by the mode of presentation of the options. So the real trouble with inconsistencies is that they foster wishful thinking and can lead to just about any conclusion – if you start out with some set of (inconsistent) information, and someone or something furnishes you with a vision to pursue that information in a certain direction, you will encounter many options along the road which you discard because they are incompatible with what you know, ending up with a firm conviction about *the* right solution, not realizing that it too is incompatible with what you know, because your 'knowledge' was inconsistent in the first place. In fact, which option you end up with all depends on which end you started to screen out, and, most likely, everyone will end up with something well in line with their initial vision. Wanton disputes will ensue, blocking co-ordinate social action. Consistency is required if you wish people to start killing their darlings and look at both, or several, sides of the matter.

ME: Badly needed, and yet impossible to attain for the human mind. Not a very neat position for someone who is hired to give me practical advice!

SS: Oh, yes, it is! I think that we are now in a good position to be constructive. We have eliminated some false starts and have also identified what problems to attack. We just have to assemble the troops. Don't you see that?

ME: No, I don't. I admit that you have relieved me of some of my preconceptions, but I don't see the constructive part.

SS: It's getting late. I suggest you take your notebook and we'll make a skeleton outline for next week. I see it like this: One often talks about people's actions as being determined by two things, what they want to achieve and what they know, values and facts. Consequently, we should affect their values or attitudes and provide them with new information if we want to influence their actions and behaviour. This is the misconception, I think. First, people very often already have the 'right' values and attitudes; I mean those that society conventionally wants its citizens to have in environmental matters. Secondly, people are quite well-informed and can seldom use much extra information in a constructive way.

What has been overlooked is that behaviour, even if ultimately based on values and information, does not spring automatically from them. Any concrete action will affect many things and therefore requires the integration of many values and much information. The missing link between values and information and actual behaviour is an integrated and coherent understanding of the greater context in which they are imbedded. The mind's limited capacity means that you can't do that from scratch in each individual case, so you have to acquire a reasonably varied stock of meaningful units – I mean chunks – in your mind. Their quality, coherence and size varies greatly depending on the capacity, efforts and interests of the individual. The main task for environmental social science is to study by what mechanisms such units are built up, maintained, and possibly reformed, and how they affect the possibility of co-ordinate social action.

ME: And do we know anything tangible about such mechanisms?

SS: Oh, yes, quite a lot. This is where our more theoretical detour returns back to firm empirical ground. You will hear about that next week.

Strategies for creating meaningful simplicity

ME: How do we organize our life in meaningful units of understanding? You promised to have an answer to that question today.

SS: I promised, not to answer it, but to show that we know quite a lot about the mechanisms behind the organization of such units. In fact, I have drawn considerably from the work of the people in *Ways Ahead*, and I would like to distinguish between three types of processes here. The first one has to do with the conscious creation of integrated mental attitudes towards some phenomenon or process, such as wind power or refuse separation, and has been studied in particular by Anna-Lisa Lindén. Under this heading we have the creation of green identities as well as commercial loyalties. The technique of selling green electricity and making people feel that they get value for their money by sending out newsletters describing the progress of renewable energy production is not significantly different from the technique of offering air travellers membership in a bonus programme with regular statements of account and special offerings. In both cases you develop a feeling of being a special kind of person, a kind you like to be and like to display in certain contexts.

ME: But this kind of attitude is focused on particular phenomena and processes. It will have a rather limited impact, will it not? Or it has to be duplicated for a great number of things.

SS: In general terms there is such a tendency, yes, but it is certainly possible to develop an identity as a person who cares for the environment in very broad terms. But the reinforcement mechanisms have to be different then, and less private – perhaps you become a member of an action group, like Greenpeace, or a green party, or you establish yourself in some other kind of organization. If this actually affects your behaviour one may say that you adopt a life style rather than an identity.

ME: How, then, can we encourage people to form green identities?

SS: First of all there must be a certain self-consciousness, a basic willingness to see oneself as someone who takes responsibility and works towards a goal. Secondly, there must be some organizer or role model to set the thing going. What we can do is to educate people to form a responsible attitude and to provide organizers. But all this has to be done at the individual level, which probably means that you have to do it differently for different people. There is a risk involved that one's devotion to a cause is vulnerable to disappointment – you enter with a vision and great expectations, and if they are not met within a reasonable period of time, you may turn your back on the whole idea. So you have to manage the thing rather carefully if you are to manage it at all.

ME: Considering this, is creation of identities a good way to go?

SS: The danger lies mainly in identities becoming too narrow. If you have to keep up involvement in too many things, it may run counter to the overall idea of mental economy.

ME: So on to your next type of process then! What is it?

SS: More automatic or habitual behaviour. It is sometimes learned but in most cases it develops spontaneously by imitation of significant others in family, school or professional life. It doesn't require an organizer or any conscious effort, and once it exists it does not require continuous reinforcement.

ME: This is the stuff that Anders Biel has studied, isn't it? Or rather, he has studied how you can *break* bad habits and create better ones, like buying eco-labeled detergents. Obviously very useful research!

SS: Certainly useful, but perhaps it is a mistake to mention the detergents as a typical case, because it gives the wrong impression about the real importance of habitual behaviour. It is the general insight that matters. People seldom realize how much of life is habitual. We may generalize from buying behaviour to car driving, refuse separation and other conventionally environment friendly behaviours, but that is not the end of it. What sort of houses we live in and how we build our cities, for what purposes we use electricity or clean water, how children should be raised – all this can in a sense be called habitual, although social conventions is a more common name for it.

I can go even further and say that the most important habits are habits of thought. The way we solve problems – you know, searching our minds for models, setting up committees, anchoring it with your superiors, and so on – how we judge people, even in the case of prejudices, and similar things, are socially determined habits of thought, and many cultural clashes are not so much caused by any deep breach between value systems as by the inability to tell a social habit from something given by nature. And the general principles for breaking or changing such social habits are still the same as those discussed by Biel.

ME: Will you give advice on how to stop religious wars as well?

SS: No, I will stop just short of that. But I do have something to say about morality – and on empirical grounds! One particularly important habit of thought is to classify things in life into different compartments with different rules: family matters are treated differently from business matters, and economic matters differently from moral matters, for example. An environmental problem can sometimes be regarded as economic and sometimes as moral, and the classification makes a difference. Biel refers to a classic study by Bruno Frey: When asked whether they would accept a nuclear waste repository in their own community, 51 per cent of the Swiss respondents answered yes. But when a scheme of monetary compensation to the inhabitants of the host community was added, the rate dropped to 25 per cent, the amount of compensation making no difference. Verbal comments by the respondents indicated that a sense of civic duty had guided the first decision – the repository had to be placed somewhere. But in the second case, the whole problem was transferred to the private economic sphere and individual incentives came to dominate. Surprising to some, but kind of expected to many social scientists.

ME: How can we encourage good habits?

SS: Formation at an early stage is much easier than reformation at a later. Develop social responsibility at an early age, use social pressure for the formation of habits and see to it that there are many visible role models in society. Start at school because many thought habits are formed there. If you need to change habits later, Biel has a good recipe.

ME: And your third kind of process?

SS: It doesn't work at the individual level but is something society does for us. I am referring to infrastructures. In a narrow sense these comprise systems for transportation, communication, supply of water and energy, collection of refuse, that is material systems for moving matter around. They have very strong connections to settlement patterns and therefore also strong environmental consequences. You are usually connected to only one road system, one water supplier and one (or very few) telephone operators, so life is much simpler than if you had to find a solution to your transportation problem, or water problem, or message problem each time anew. Arne Kaijser, in his chapter in this book, summarizes the success of infrasystems in three words: cheap, convenient and reliable, but also points out that it is exactly these properties that cause environmental problems: it becomes so easy to overuse resources. To this can be added that the large-scale standardized operation of infrasystems controls the behaviour of individuals in many ways and leaves little room for individual incentives to change that behaviour in a more environment friendly way.

In a broader sense, infrasystems include also social structures, necessary for their running and maintenance, as Kaijser has repeatedly pointed out. They give rise to standardizations, both within the systems (like gauge or voltage) and in a more general sense of promoting unified systems of measurement. Perhaps the concept of an infrasystem can be extended to include also such purely immaterial systems as the metre system, probably the most important infrasystem of all, although the invention of the coin and the concept of currency by king Croesus of Lydia is also a good candidate. In the long run the important thing about power lines may not be their geographical positions but the fact that they have forced standardization of voltage and periodicity

ME: What's important for me in this? How does it affect the environment?

SS: What I mean to point out is that infrasystems have strong long term effects. If they were not designed with the environment in mind, it may well be that they don't affect the environment very much or do it in a harmful way, but if you *do* design them with the environment in mind, they may constitute very powerful means for improving it. They are an important strategic tool. To parallel the power lines with something much simpler: perhaps the lasting impact of refuse collecting systems is that they force us to use new types of bins which make it very impractical *not* to separate refuse.

ME: You emphasize human weakness more than human strength, you look upon human beings as creatures led by the law of least resistance, and you say that we should organize society so that people are guided by the system and don't have to make their own decisions. I think you reduce human beings to brute beasts. Don't you think we should pay any attention at all to human idealism, to people's dedication to the environment, to a sense of responsibility for other people and for future generations?

SS: Pay attention – yes, rest satisfied with – no! I think many underestimate the scale of the problem. What we need to accomplish are large and persisting effects. Perhaps you can, by keeping the environment always on the agenda, keep many people dedicated some of the time and a few most of the time, but you will never be able to keep most of the people dedicated to the environment most of the time, no matter how much effort you put in. And we don't really want it either – there must be some mental energy left in society for other causes: peace, social justice and health merit some attention too!

ME: Don't you think we need charismatic figure-heads?

SS: I'm a bit uncertain. We certainly need to have people with strong commitment and a thorough knowledge of the environment in those positions where the social systems are formed, I mean in politics, in government agencies, and in trade and industry, and we need to maintain a mental atmosphere in those circles which makes these people and their work respected. But they don't have to be charismatic; it is much more important that they are knowledgeable and efficient.

ME: But in public life?

SS: Charisma is a great help in public life, and society probably needs role models for environmental behaviour. But I have somewhat heretic views about this, and I see a risk with strong devotion to the environment. Recall the 'chunks' we spoke about earlier and how difficult it is for people to be globally consistent about things which have entered their mind by way of different chunks. The disciples of a very charismatic leader often become dedicated to very narrow problems, they next to perfect the internal logic of a given chunk at the expense of all other chunks. What they call eco-fascism is an extreme example at a high level, but you also see the same thing at lower levels: wind power or refuse separation become ends in themselves, no matter how much back-up effect one must build or how much transport it generates. A moderate variety is Mona Mårtensson's observation, which I mentioned at our second meeting, about people who have chosen to live in eco-villages and conscientiously stick to certain kinds of environment friendly behaviour, but who nevertheless contribute to the overall environmental load as much as anyone else. You may call it by various names – mental blinkers or tunnel-vision, for example – but it is a common enough phenomenon.

ME: Now, let me try to sort all this out. I called on you because I had a problem: I was very sceptical about the value of the social sciences for solving environmental problems – not because I didn't see the importance of social causes, but because the results were coming too slowly to be of any real use. You have tried – and very eloquently – to persuade me of the opposite. But how far have we come? Is there any environmental problem about which you have taught me enough so that I can say: *this* is the real environmental problem and *this* is how we shall solve it?

SS: Perhaps not, but for good reasons. What I *have* shown is that you make an unwarranted presupposition in the very formulation of your question. There are no environmental problems! At least not of the kind where the social sciences are involved. But there are many important and urgent problems ranging over several sectors of society, including the environment. The environmental aspect may be stronger or weaker, but is seldom dominating – few problems can be solved on environmental grounds only. If you separate out a particular problem and label it environmental, you pave the way for limited local consistency and for conflicts with other values. People will feel very self-satisfied when they have done what is in fact only a fragmentary contribution. The solution is to arrange the pay-off structure for *each* problem so that greater weight is put on the environmental aspect, rather than picking out some problems as specifically environmental.

ME: Which are these multi-range problems, then?

SS: All the usual ones, the ones we have always discussed but perhaps not from the point of view of the environment. Urban planning, transportation systems, and energy supply are obvious ones, and everybody is now aware that they have important social consequences. But there are also more basic factors, originating in the social sphere, like mobility and migration, cohabitation patterns, nativity, and perhaps 'life style'. And there are mental and ideological factors, like degree of liberalism, type of democracy, and citizens' general attitude towards rights and duties which form a framework for everything else.

All these problems are right there, before our eyes, and they all have consequences for the environment. We must first and foremost identify all these consequences and the great challenge is then to learn to understand how the different problem spheres interact. We must aim for global consistency, and what we above all need is good theoretical research to make us able to separate the fundamental from the superficial and to reduce the bewildering multitude of consequences to a manageable number of principles.

ME: Again, but from different premises, a plea for more fundamental research. Am I right?

SS: Yes, you are.

ME: You have made your point very clear, and you have convinced me that the social sciences have a role to play. But I haven't become much clearer about what I should do to promote the right kind of social science.

SS: Oh, then you must ...

ME: No thank you! I realize that you must have views on that problem too, but it was not part of your task, and it will have to wait for another occasion. Goodbye, and many thanks for what you've done!

End note

I have been dropping names here and there, but I have drawn on the work of my fellow researchers in *Ways Ahead* in innumerable other places as well, leaving out the references for purely stylistic reasons.

Social-Ecological Resilience and Behavioural Responses

Carl Folke

Introduction

A common thread of this volume is the analysis of behavioural responses to environmental issues and environmental change; how socially determined habits of thought and action, from the level of the individual to society as a whole, govern environmental behaviour. A key challenge is to unravel mechanisms behind the creation of habits of thought and action and how they are maintained and possibly reformed, recognizing that behavioural changes have multiple determinants. There is an overarching emphasis on the processes of behavioural change and their dynamic and non-linear nature.

The focus of the volume provides a refreshing contrast to the simplified view and application of individual rationality in for example social cost/benefit analysis (e.g. Pearce et al., 1989), an analysis often promoted as an essential tool to capture the social significance of the environment and help make informed and rational decisions. Such an analysis does not take into account the inherent complexities and resulting uncertainties associated with dynamic and interdependent human-environmental systems (Pritchard et al., 2001). Preferences and values of people are not necessarily invariable, nor do they exist in a social and cultural vacuum; rather, they are formed and re-formed as part of a social process, part deliberative, part historical (North, 1990). They co-evolve with a diversity of social and environmental variables over a diversity of temporal and spatial scales with processes that seldom operate in a smooth fashion. Furthermore values and behaviours do not necessarily coincide. There are conflicts and mismatches between values for the environment and human behaviour and there is also inertia to behavioural change (Biel, chapter 2; Mårtensson and Pettersson, chapter 3). People may be well informed and concerned about environmental issues but the social and institutional context in which they are embedded may offset behavioural responses. For example, trajectories of large technical systems, or socially constructed infrasystems, are difficult to redirect. Infrasystems leave vast and long-lasting imprints on society and constrain behavioural change (Kaijser, chapter 7).

But there are also contexts that stimulate behavioural responses such as the interplay between consumers and producers in creating green identities and products (Lindén and Klintman, chapter 4), the environmental repositioning of large firms to fit the sustainability discourse in society (Wolff and Zaring, chapter 5), the self-organizing responses of the global business community towards sustainable management practices and social responsibility (Hydén and Gillberg, chapter 6), lurches of social learning (Lee, 1993) and formation of norms and rules for collective action to cope with environmental change (Ostrom, 1990; Berkes and Folke, 1998). Thus, directing human behaviour towards improved environmental performance and sustainability is not just a simple matter of providing information and policy prescriptions but a complex socio-cultural process. It will require understanding of the contexts that form, shape and reshape habits of thought and action. As stated by Hansson (chapter 8) 'the missing link between values and information and actual behaviour is an integrated and coherent understanding of the greater context in which they are imbedded'.

The authors of the volume are confronted with 'the greater context' from multiple perspectives and at different temporal and spatial scales. Some are faced with contexts that constrain behavioural change; other chapters concern contexts that support such change. Empirical interdisciplinary work crossing the boundaries of the natural and social sciences, inspired by the theories of C.S. Holling, suggest that change in environmental behaviour and resource management is less likely to take place during periods of growth and stable conditions. It is during periods of rapid change (or release in Holling's terminology), often perceived as periods of crisis, and in the following reorganization that renewal and redirection of social pattern and behaviour are most likely to happen (Holling, 1986; Holling and Sanderson, 1996; Holling, 2001). Such periods open up space for transformations, from one behaviour to another, from one perspective to another. It is a time of crisis, but also of opportunity framed by previous experience and social memory of the system (McIntosh, 2000; Folke et al., 2003).[1] There is a story about two persons who were fired during a period of rapid economic recession. By chance they happened to be in the same inn, sitting at the same table, during this period of personal strain and corporate crises. They had never met before, but started to talk, developed friendship and trust, and discovered the potential in combining their previous experiences into new products of the sport shoe industry, a business that

---

[1]Social memory is the arena in which captured experience with change and successful adaptations, embedded in a deeper level of values is actualized through community debate and decision-making processes into appropriate strategies for dealing with ongoing change (McIntosh, 2000).

turned out to be very successful. Unexpected interactions like those can occur among previously separate properties in nature and society that can nucleate an inherently novel and unexpected focus for future good or ill.

Understanding the contexts for how people respond to and shape periods of change and how society reorganizes following change seems to be a largely neglected and poorly understood issue in environmental science and management (Gunderson and Holling, 2002). It will most likely require a stronger emphasis on thinking that moves from the perspective of a world in steady state or near-equilibrium to one of complex systems (Holland, 1995; Kauffman, 1993).

Assessing and evaluating sustainability in the context of complex systems is considered a frontier of interdisciplinary research (Ludwig et al., 2001). New perspectives, concepts and tools about the dynamics of complex systems and their implications for sustainability are now developing in parallel, influencing the natural sciences, the social sciences and the humanities through the work of many people and groups. Complex systems thinking is for example used to bridge social and biophysical sciences to understand climate, history and human action (McIntosh et al., 2000), assessments of regions at risk (Kasperson et al., 1995), syndromes of global change (Petschel-Held et al., 1999) and how to link social and ecological systems for sustainability (Berkes and Folke, 1998; Gunderson and Holling, 2002; Berkes et al., 2003). It underpins many of the new integrative approaches such as ecological economics (Costanza et al., 1993; Costanza et al., 2001; Arrow et al., 1995) and sustainability science (Kates et al., 2001; Clark et al., 2001). A long-term perspective suggests that stability in the management of complex systems is an illusion that disappears when one chooses a scale of perception commensurate with the phenomena under investigation (van der Leeuw, 2000). A long view also highlights the importance of scale interactions across time and space in relation to adaptive renewal cycles of growth, conservation, release and reorganization in social and ecological systems (Gunderson and Holling, 2002).

In the remaining part of the chapter I intend to address social responses to environmental change, and in particular to changes in resource and ecosystem dynamics. I will do this using the concept of resilience as a framework for the discussion. Resilience provides the capacity to absorb sudden change, cope with uncertainty and surprises while maintaining desirable functions. Resilience provides the components for renewal and reorganization following change. Vulnerability is the flip side of resilience: when a social or ecological system loses resilience it becomes vulnerable to change that previously could be absorbed. In a resilient system, change has the potential to create opportunity for development, novelty and innovation. In a vulnerable system even small changes may be devastating. The concept of resilience shifts perspective from the aspiration to control change in systems assumed to be stable, to sustain and enhance the capacity of social-ecological systems to cope with, adapt to, and shape change. The degree to which the social-ecological system can build and increase the capacity for learning, adaptation and responding in a manner that does not constrain or erode future opportunities is a central aspect of resilience (Carpenter et al., 2001; Berkes et al., 2003). The use of the resilience concept in the social sciences is reviewed by e.g. Davidson-Hunt and Berkes (2003), Scoones (1999) and King (1995). Social resilience is discussed in e.g. Adger (2000).

The first part of the chapter gives a brief background of interactions and interdependencies between the life-supporting environment and societal development. This background provides an entry into a synthesis of a common pattern of response to environmental change and its implication for sustainability that has recently emerged from both contemporary and historical interdisciplinary studies of resource management systems. The synthesis exposes how socially determined habits of thought and action reinforced by short-term successes have led to ecological and social vulnerability in the longer term. It also highlights that if sustainability is the desirable direction for societal development it will not be sufficient to build an integrated and coherent understanding of only the social context in which environmental behaviours are embedded. The development of an integrated view of coupled social and ecological systems and a coherent understanding of social-ecological contexts need to be nurtured. A few modest attempts in this direction are presented.

The Life-Support Environment and Societal Development

Earth's life-support systems do not develop in a smooth deterministic fashion. They are complex systems with non-linearities, thresholds and multiple stability domains (Levin, 1999). A bundle of disturbances at different temporal and spatial scales – a disturbance regime – are part of ecosystem dynamics and development. The disturbance regime contributes to building healthy ecosystems. Disturbance opens up patches of opportunity for renewal and reorganization of the ecosystem, for development and evolution. Several studies have shown how increasingly nested human activities in the biosphere are changing disturbance regimes by:
1. actively suppressing or removing disturbance;
2. transforming pulse events into persistent disturbance or even chronic stress;
3. by introducing new disturbances.

The intensity, severity, duration, spatial distribution, and frequency of disturbances are altered. Combinations of those changes lead to new synergistic effects termed compounded perturbations that in many aspects are new to organisms and ecosystem dynamics (Paine et al., 1998; Nyström and Folke, 2001).

Ecological resilience, the capacity to buffer or absorb disturbance, is required for reorganization following change (Holling, 1986). Ecological resilience contributes in time and space with the network of species, their dynamic interactions between each other and the environment, and the combination of structures that make reorganization after disturbance possible. Hence, resilience is a key property of the life-support environment. It sustains a flow of essential ecosystem services on which social and economic development depends through the dynamic capacity to absorb disturbance and provides the components for reorganization, opportunity and novelty.

Throughout history humanity has shaped nature and nature has shaped the development of human society. The main part of Earth's surface has been modified by human activities (Turner et al., 1990) and recently at a much faster pace than earlier in human history. There are neither natural or pristine systems, nor are there social systems without nature. Instead humanity and nature have been co-evolving within the biosphere in a dynamic fashion (Norgaard, 1994) and will continue to do so. Human actions are a major structuring factor of the life-supporting environment, and despite tremendous improvements in technological, economic and material well being, in some parts of the world, development of human society in all parts of the world will continue to rely on the capacity of the biosphere to provide ecosystems services and support.

Throughout human history there has been a tendency to homogenize ecosystems for production of certain valuable resources (Redman, 1999) a tendency that has escalated since the Second World War. During the last century the human population increased by a factor 4, the urban population by a factor 13, water use 9, sulphur dioxide emissions 13, carbon dioxide emissions 17, marine fish catch 35 and industrial output 40 times (McNeill, 2000). With this expansion there has been a major reshuffling in land cover towards simplified landscapes for production of food, timber and other renewable resources. Homogenization causes loss of resilience. Ecosystems with reduced resilience may still maintain function and generate services, i.e. may seem to be in good shape. But when faced with an additional disturbance a critical threshold may be reached as a consequence of loss of resilience, and the system may slide into an undesirable stability domain where a large-scale degradation may occur, a pattern observed in both terrestrial and aquatic ecosystems (Scheffer at al., 2001).

A disturbance that earlier triggered a dynamic development of the system may under circumstances of lowered resilience become an obstacle to development. Losses of resilience through impacts on the landscape and seascape will exacerbate the effects of changed disturbance regimes and compounded perturbations and increase the likelihood for shifts into socially undesirable stability domains (Nyström et al., 2000; Jackson et al., 2001). These shifts are sometimes irreversible and in other cases the costs (in time and resources) of reversal are so large that reversal may be impractical, as illustrated for lakes when they shift from clear water to turbid water or rangelands when they shift from productive grasslands to shrub landscapes (Carpenter et al., 2001). Such shifts may narrow the potential for social and economic development, reduce options for livelihoods, and create environmental refugees as a consequence of the impact on ecosystem life-support.

Hence, the likelihood of rapid environmental transformations and ecological surprises increases with altered disturbance regimes and reduced resilience. Or put in other words – social vulnerability is likely to increase and opportunity for development is likely to be constrained if society erodes resilience (Folke et al., 2002).

The Regional Pathology of Resource Management

Human simplification of landscapes and seascapes for production of particular target resources to be traded on markets has generated steady resource flows in the short term. But it has done so at the expense of reduced diversity and it has eroded resilience. Far too often managers seek to command-and-control processes of change for optimal production in simplified landscapes in an attempt to stabilize resource outputs and sustain consumption patterns. Short-term successes of increasing yield in homogenized environments seem to reinforce a social perception of humanity as superior to and independent of nature. Nature can be conquered, controlled and ruled. Further efforts are made to reduce environmental variability and remove disturbance. The life-supporting environment is transformed into an economic sector for production of social value. Short-term successes make managing ecosystem dynamics a marginal issue and as a consequence knowledge, incentives and institutions for monitoring and responding to environmental feedback erode. Short-term successes cause managers to shift their attention from the original purpose of sound resource management to efforts to increase organizational or economic efficiency (Gunderson et al., 1995).

Since disturbance is endogenous to the cyclic processes of ecosystem renewal, from local scales to the biosphere, this type of resource management tends to increase the potential for larger-scale disturbances and even less predictable and less manageable feedbacks, or surprises, from the environment (Gunderson and Holling, 2002). Technological systems, or infrasystems, may further mask the feedback from the environment, thereby magnifying the accumulation of disturbance to larger spatial and longer temporal scales (Kaijser, chapter 7). The behaviour unconsciously contributes to a modification of the important variables that structure and sustain desirable states (Carpenter et al., 2001). Society becomes more susceptible to surprise and crisis but is ignorant about it. Vulnerability is created without recognizing it (Kasperson et al., 1995). An example is the cod resource collapse in Newfoundland, which had been predicted by inshore fishers and some field biologists (Finlayson and McCay, 1998). The problem was exacerbated by 'an overreliance on the science and culture of quantitative stock assessment' (National Research Council, 1998: 35) by central government agency population modellers who (in retrospect) misused or misjudged their data, and precipitated a stock collapse unprecedented in its magnitude in the North Atlantic.

This pattern of environmental management, briefly summarized and simplified above, has been termed the 'pathology of natural resource management' (Holling and Meffe, 1996) and has been described for several sectors, in several regions of the world and over different temporal scales (e.g. Regier and Baskerville, 1986; Gunderson et al., 1995: Redman, 1999; Carpenter and Gunderson, 2001). According to Holling (2003) the regional pathology has the following features:

1. The policies and development initially succeed in removing disturbance and enhancing growth.
2. Implementing agencies initially are responsive to the ecological, economic and social forces, but evolve to become narrow, rigid and myopic. They become captured by economic dependents and the perceived needs for their own survival.
3. Economic sectors affected by the resources grow and become increasingly dependent on perverse subsidies.
4. The relevant ecosystems gradually lose resilience to become fragile and vulnerable and more homogeneous as diversity and spatial variability is reduced.
5. Crises and vulnerabilities begin to become more likely and evident and the public begins to loose trust in governance.

In rich regions the resulting crises have led to spasmodic lurches of learning with expensive actions directed to reverse the worst of the consequences of past mistakes. In poor regions the result has been dislocation of people, increasing uncertainty, impoverishment and a poverty trap.

van der Leeuw (2000) characterizes land degradation and the creation of vulnerability as a socio-natural process that has occurred throughout history, a process that highlights the importance of the underlying perception of the socio-natural system. Human drivers of ecosystem change are deeply embedded in cultural values and underlying perceptions (Thompson et al., 1990), and economic production systems and lifestyles, mediated by institutional factors (Lambin et al., 2001). Urbanization and many aspects of globalization tend to distance people from their relation to ecosystem support by disconnecting production from consumption and production of knowledge from its application (Folke et al., 1998). People become alienated both physically and mentally from their dependence on access to resources and ecosystem functions outside the boundaries of their own jurisdiction.

Facing complex co-evolving social-ecological systems for sustainability requires ability to cope with, adapt to and shape change without losing options for future adaptability. It is not about controlling or removing change. The paradox is that the mental model of optimal management of systems assumed to be stable and predictable has in many respects reduced the potential for development and altered the capacity of life-support ecosystems to buffer change. The less resilient the system, the lower is the capacity of institutions and societies to adapt to and shape change. Managing for resilience is therefore not only an issue of sustaining capacity and opportunity for development, now and in the future, but also an issue of environmental, social and economic security (Adger et al., 2001).

Behavioural Responses for Social-Ecological Resilience

Obviously, human environmental responses are more diverse and multiple than sketched above. There are places and societies that practiced sustainable resource use, not merely of resources but entire ecosystems, and even whole drainage basins, and some of their adaptations survive to date (Gadgil et al., 1993; Berkes and Folke, 1998). In these societies a pattern of co-evolutionary adaptations between social systems and natural systems must have been the norm (Norgaard, 1994), with the adaptations in many cases driven by crises, learning and redesign. Individual preferences seem to have acted in a social context that promoted sustainability of the combined and co-evolving social-ecological system, simply because behaving in a sustainable fashion was a necessity for survival. The co-evolutionary character reflects the fact that social-ecological systems can change qualitatively to generate and implement innovations that are truly creative, in the sense of opportunities for novel cooperation and feedback management.

Some of the most sophisticated co-evolving systems are common-property institutions that have developed over long periods of time (Ostrom et al., 1999). Examples include Spanish *huertas* for irrigation, Swiss grazing commons (Ostrom, 1990) and marine resource tenure systems in Oceania (Ruddle et al., 1992). In other areas, such institutions have evolved over a short period of time (in the order of one decade) in response to a management crisis. Examples include the Turkish Mediterranean coastal fishery in Alanya (Berkes, 1992) or the watershed-based resource management system in western Sweden (Olsson and Folke, 2001). There seem to be social mechanisms in place that respond to ecological feedbacks instead of blocking them out (Berkes and Folke, 1998).

Are there behavioural responses that sustain social-ecological systems in a world that is constantly changing? Such issues are addressed through a case study approach in a recent volume (Berkes et al., 2003) focusing on periods of change caused by disturbance, surprise or crisis, followed by renewal and reorganization. Folke et al. (2003) identify and expand on four critical factors highlighted in many of the chapters of the volume, behavioural responses that interact across temporal and spatial scales and that seem to be required for dealing with resource dynamics in social-ecological systems:

- learning to live with change and uncertainty;
- nurturing diversity for reorganization and renewal;
- combining different types of knowledge for learning;
- creating opportunity for self-organization towards social-ecological sustainability.

Learning to Live with Change and Uncertainty

The first factor emphasizes the necessity of accepting change and living with uncertainty and surprise, and the Berkes et al. volume provides examples of strategies of social-ecological management that takes advantage of change and crisis and turns it into opportunity for development. Management that actively behaves like disturbance is one of a sequence of practices – ecological and social – that seems to generate resilience (Berkes and Folke, 2002). Examples include small-scale burning of forests and pastures for pest control and pulses of herbivore grazing as practiced by many local communities worldwide. It appreciates the role of disturbance in development and includes monitoring and ecological knowledge and understanding of ecosystem condition and dynamics embedded in social institutions. Such management practices seem to have developed as a result of actual experience with change and crisis, realizing that not all possible outcomes can be predicted and planned for. Responding based on such experience depends on institutional learning incorporating previous crises, and may help avoid unwanted qualitative shifts in stability domains of resource systems. In this sense, institutions emerge as a response to crisis and are reshaped by crisis (Olsson and Folke, 2001). Several of these local resource and ecosystem management strategies and associated institutions presented in the Berkes et al. volume resemble risk spreading and insurance building within society, similar to portfolio management in financial markets (Costanza et al., 2000). As suggested by Low et al. (2003) diversity and redundancy of institutions and their overlapping functions may play a central role in absorbing disturbance and in spreading risks, just like diversity and redundancy of species and their function in ecosystem resilience.

Nurturing Diversity for Resilience

The second factor illuminates the importance of nurturing diversity for resilience, recognizing that diversity is more than insurance to uncertainty and surprise. It also provides the bundle of components, and their history, that makes development and innovation following disturbance and crisis possible, components that are embedded in the social memory of ecosystem management (Folke et al., 2003). Hence, diversity also plays an important role in the reorganization and renewal process following disturbance. Folke et al. (2003) hypothesize that it is in this context that the social memory becomes significant, because it provides a framework of accumulated experience for coping with change.

The results of the Berkes et al. volume suggest that the experience of the role of disturbance, uncertainty and surprise, and the need to nurture biodiversity and conserve ecological memory for maintaining adaptive capacity, must be stored in the social memory of resource users and managers and be expressed in practices that build resilience. These include conflict resolution, negotiation, participation and other mechanisms for collaboration with rules aimed at maintaining the process of learning and adaptation in situations facing uncertainty and change. It also seems to require a nested social network, operating at multiple scales, with trust and respect for ecosystem management.

Combining Knowledge Systems into Institutions

The third factor of environmental response addresses the significance of peoples' knowledge, experience and understanding about the dynamics of complex ecosystems, their inclusion in management institutions, and their complementarities to conventional management. An important aspect is that an adaptive learning process for managing ecosystems for social-ecological resilience should not dilute, homogenize, or diminish the diversity of experiential knowledge systems for ecosystem management, since they may embed lessons for how to respond to change and how to nurture diversity. Scientific understandings of complex adaptive systems and their change could be enriched by insights from local ecosystem management. There is also a need to expand knowledge from structure of nature to function of nature when dealing with complex systems. The potential for learning and building social-ecological resilience by making use of and combining different knowledge systems should be taken seriously.

Furthermore, the significance of incorporating knowledge of ecological processes and dynamics into institutions needs to be recognized. Knowledge acquisition is an ongoing dynamic learning process; perhaps most importantly, it seems to require social networks and an institutional framework to be effective. Flexible social networks and organizations that proceed through learning-by-doing are better adapted for long-term survival than are rigid social systems that have set prescriptions for resource use. Such flexible institutional arrangements have been judged as inefficient since they look messy and are non-hierarchical in structure. A growing literature on polycentric institutions (McGinnis, 2000) is demonstrating that dynamic efficiency is frequently thwarted by creating centralized institutions and enhanced by systems of governance that exist at multiple levels with some degree of autonomy complemented by modest overlaps in authority and capability. A diversified decision-making structure allows for testing of rules at different scales and contributes to the creation of an institutional dynamics important in resilience management.

Thus, it is not effective to separate ecological studies aimed at management from the institutional framework within which management takes place. Understanding ecosystem processes and managing them is a progression of social-ecological co-evolution, and it requires learning and accumulation of ecological knowledge and understanding in the social memory (McIntosh, 2000). In that sense, a collective learning process, that builds knowledge and experience with ecosystem change, evolves as a part of the institutional and social memory, and it embeds practices that nurture ecological memory.

Creating Opportunity for Self-Organization

The fourth factor brings together the behavioural responses above in the context of self-organization, including scale, governance and external drivers, and emphasizes the significance of the dynamic interplay between diversity and disturbance (Folke et al., 2003). Both diversity and disturbance are parts of sustainable development and resilience and their interaction needs to be explicitly accounted for in an increasingly globalized and human dominated biosphere (Gunderson and Holling, 2002).

The learning process is of central importance for social-ecological capacity to build resilience. It is important that learning processes include operational monitoring and evaluation mechanisms in order to generate and refine ecological knowledge and understanding into management institutions. This is the focus of adaptive co-management in which institutional arrangements and ecological knowledge are tested in an ongoing trial-and-error process. Adaptive co-management draws on social-ecological memory and is informed by both practice and theory. It relies on the participation of a diverse set of interest groups operating at different scales. However, creating platforms for conflict resolution and participation by various interest groups for learning and knowledge creation will not be sufficient for sustainability. It requires the context of the dynamic interplay between diversity and disturbance in resilience. Ecological knowledge and understanding of this interplay is a necessity and social-ecological memory frames the process.

It has been suggested that diversity in functions and in response among local level resource management systems, from the individual level to organizations and institutions (Burger et al., 2001; Westley, 2002), enhances performance so long as there are overlapping units of government that can resolve conflicts, aggregate knowledge across scale, and insure that when problems occur in smaller units, a larger unit can temporarily step in (Low et al., 2003). Cash and Moser (2000) propose that governance for linking global and local scales should utilize boundary organizations, utilize scale-dependent comparative advantages, and employ adaptive assessment and management strategies. Governments should nurture the self-organizing ability of actors to voluntarily develop new norms and codes of conduct for sustainability (Hydén and Gillberg, chapter 6) and create space for flexible and innovative collaboration towards sustainability (Folke et al., 2002).

Multi-level governance of complex ecosystems needs constant adjustment, which requires innovation and experimentation (Shannon and Antypas, 1997; Imperial, 1999). Olsson and Folke (2001) describe the development of watershed management by a local fishing association in a multi-level governance system faced with internal and external ecological and social change. The social change included devolution of management rights, which provided an arena for local users to self-organize and developed, refine, and implement rules for ecosystem management. Not only do these people respond to change but by doing so they learn, develop a social memory and build adaptive capacity to deal with future change in the multi-level governance system. Working with such 'open institutions' is essential for dealing with ambiguity of multiple objectives, uncertainty and the possibility of surprising outcomes (Shannon and Antypas, 1997; Kasperson and Kasperson, 2001). Such emergent governance (Shannon, SUNY Buffalo Law School, pers.comm.) that creates new institutional platforms for adaptive management is evolving in many places.

Concluding Remarks

The focus of the volume has been to understand social contexts underlying behavioural responses to environmental issues and environmental change. The chapters have illustrated that those contexts can both constrain and support responses toward environmental improvement. It has been suggested that crises may play a constructive role in opening up space for reshaping of habits of thought and action, for removal of barriers to behavioural responses redirecting societal development towards sustainability. I have argued that if sustainability is a societal goal, than those social contexts have to be linked to an appreciation of the co-evolving nature of societal development and biospheric processes; the dynamic interplay that takes place from local to global scales. Social constructs that disconnect society and nature and that treat the environment as a separate sector have led to regional development pathologies and caused vulnerability.

Building social-ecological resilience for sustainability requires a fundamental shift in thinking and perspective from social constructs that assume a world in steady-state to be controlled by focusing on preventing and inhibiting change, to a deeper understanding of contexts that recognize change as the rule rather than the exception, and concerned with the capacity of complex adaptive social-ecological systems to live with change and shape change without constraining future options for development.

References

Adger, W. N., (2000), 'Social and ecological resilience: are they related?', *Progress in Human Geography*, 24:347–364.

Adger, W. N., Neil, P., Kelly, M., and Huu Ninh, N., (2001), *Living with Environmental Change: Social Vulnerability, Adaptation, and Resilience in Vietnam*, Routledge, London.

Arrow, K., Bolin, B., Costanza, R., Dasgupta, P., Folke, C., Holling, C. S., Jansson, B.-O., Levin, S., Mäler, K.-G., Perrings, C., and Pimentel, D., (1995), 'Economic growth, carrying capacity and the environment', *Science*, 268: 520–521.

Berkes, F., (1992), 'Success and failure in marine coastal fisheries of Turkey', in D. W. Bromley (ed.), *Making the Commons Work. Institute for Contemporary Studies*, San Francisco: 161–182.

Berkes, F., and Folke, C., (eds) (1998), *Linking social and ecological systems: management practices and social mechanisms for building resilience*, Cambridge, UK: Cambridge University Press.

Berkes, F., and Folke, C., (2002), 'Back to the future: Ecosystem dynamics and local knowledge', in L. H. Gunderson and C. S. Holling (eds) *Panarchy: Understanding Transformations in Systems of Humans and Nature*, Island Press, Washington DC.

Berkes, F., Colding, J., and Folke, C., (eds) (2003), *Navigating Social-Ecological Systems: Building Resilience for Complexity and Change*, Cambridge University Press, Cambridge, UK.

Burger, J., Ostrom, E., Norgaard, R. B., Policansky, D., and Goldstein, B. D., (2001), *Protecting the Commons: A Framework for Resource Management in the Americas*, Island Press, Washington, DC.

Carpenter, S. R., and Gunderson, L. H., (2001), 'Coping with collapse: Ecological and social dynamics in ecosystem management', *BioScience*, 51:451–457.

Carpenter, S. R., Walker, B., Anderies, J. M., and Abel, N., (2001), 'From metaphor to measurement: Resilience of what to what?', *Ecosystems*, 4:765–781.

Cash, D. W., and Moser, S. C., (2000), 'Linking global and local scales: Designing dynamic assessment and management processes', *Global Environmental Change*, 10:109–120.

Clark, W., Jäger, J., van Eijndhoven, J., and Dickson, N., (eds) (2001), *Learning to Manage Global Environmental Risks: A Comparative History of Social Responses to Climate Change, Ozone Depletion, and Acid Rain*, The Social Learning Group, MIT Press, Cambridge, MA.

Costanza, R., Daly, H., Folke, C., Hawken, P., Holling, C. S., McMichael, A. J., Pimentel, D., and Rapport, D., (2000), 'Managing our environmental portfolio', *BioScience*, 50:149–155.

Costanza, R., Low, B. S., Ostrom, E., and Wilson, J., (2001), *Institutions, Ecosystems, and Sustainability*, Lewis Publishers, Boca Raton.

Costanza, R., Waigner, L., Folke, C., and Mäler, K.-G., (1993), 'Modeling complex ecological economic systems: toward an evolutionary dynamic understanding of people and nature', *BioScience*, 43:545–555.

Davidson-Hunt, I. J., and Berkes, F., (2003), 'Nature and Society through the Lens of Resilience: Toward a Human-in-Ecosystem Perspective', in F. Berkes, J. Colding and C. Folke (eds), *Navigating Social-Ecological Systems: Building Resilience for Complexity and Change*, Cambridge University Press, Cambridge.

Finlayson, A. C., and. McCay, B. J., (1998) 'Crossing the threshold of ecosystem resilience: the commercial extinction of northern cod', in F. Berkes and C. Folke (eds), *Linking Social and Ecological Systems: Management Practices and Social Mechanisms for Building Resilience*, pp. 311–337, Cambridge: Cambridge University Press.

Folke, C., Carpenter, S., Elmqvist, T., Gunderson, L., Holling, C. S., Walker, B., Bengtsson, J., Berkes, F., Colding, J., Danell, K., Falkenmark, M., Moberg, M., Gordon, L., Kaspersson, R., Kautsky, N., Kinzig, A., Levin, S. A., Mäler, K.-G., Ohlsson, L., Olsson, P., Ostrom, E., Reid, W., Rockström, J., Savenije, S., and Svedin, U., (2002), *Resilience and Sustainable Development: Building Adaptive Capacity in a World of Transformations*, ICSU Series on Science for Sustainable Development No. 3, 2002, International Council for Science, Paris and Report for The Swedish Environmental Advisory Council No 1. Ministry of the Environment, Stockholm.

Folke, C., Colding, J., and Berkes, F., (2003), 'Building resilience for adaptive capacity in social-ecological systems', in F. Berkes, J. Colding and C. Folke (eds), *Navigating Social-Ecological Systems: Building Resilience for Complexity and Change*, Cambridge University Press, Cambridge.

Folke, C., Pritchard Jr., L., Berkes, F., Colding, J., and Svedin, U., (1998), *The problem of fit between ecosystems and institutions*, IHDP Working Papers No. 2, International Human Dimensions Program on Global Environmental Change.

Gadgil, M., Berkes, F., and Folke, C., (1993), 'Indigenous knowledge for biodiversity conservation', *Ambio*, 22: 151–156.

Gunderson, L. H., and Holling, C. S., (eds) (2002), *Panarchy: Understanding Transformations in Systems of Humans and Nature*, Island Press, Washington, DC.

Gunderson, L. H., Holling, C. S., and Light, S., (eds) (1995), *Barriers and Bridges to the Renewal of Ecosystems and Institutions*, Columbia University Press, New York, NY.

Holland, J. H., (1995), *Hidden Order: How Adaptation Builds Complexity*, Addison-Westley, Reading, MA.

Holling, C. S., (1986), 'The resilience of terrestrial ecosystems: local surprise and global change', in W. C. Clark, R. E. Munn, (eds), *Sustainable Development of the Biosphere*, International Institute for Applied Systems Analysis, (IIASA), Cambridge University Press, Cambridge, UK.

Holling, C. S., (2001), 'Understanding the complexity of economic, ecological and social systems', *Ecosystems*, 4: 390–405.

Holling, C. S., (2003), 'The back-loop to sustainability', in F. Berkes, J. Colding and C. Folke (eds), *Navigating Social-Ecological Systems: Building Resilience for Complexity and Change*, Cambridge University Press, Cambridge.

Holling, C. S., and Meffe, G. K., (1996), 'Command and control and the pathology of natural resource management', *Conservation Biology*, 10:328–37.

Holling, C. S., and Sanderson, S., (1996), 'Dynamics of (dis)harmony in ecological and social systems', in S. Hanna, C. Folke and K.-G. Mäler, (eds), *Rights to Nature: Ecological, Economic, Cultural, and Political Principles of Institutions for the Environment*, Island Press, Washington, DC.

Imperial, M. T., (1999), 'Institutional analysis and ecosystem-based management: the institutional analysis and development framework', *Environmental Management*, 24:449–65.

Jackson, B. C., Kirby, M. X., Berger, W. H., et al. (2001), 'Historical overfishing and the recent collapse of coastal ecosystems', *Science*, 293:629–638.

Kasperson, J. X., and Kasperson, R. E., (eds) (2001), *Global Environmental Risk*, United Nations University Press/Earthscan, London.

Kasperson, J. X., Kasperson, R. E., and Turner, B. L., (1995), *Regions at Risk: Comparisons of Threatened Environments*, United Nations University Press, NY.

Kates, R. W., Clark, W. C., Corell, R., Hall, J. M., Jaeger, C. C., Lowe, I., McCarthy, J. J., Schellhuber, H. J., Bolin, B., Dickson, N. M., Faucheux, S., Gallopin, G. C., Grubler, A., Huntley, B., Jäger, J., Jodha, N. S., Kasperson, R. E., Mabogunje, A., Matson, P., Mooney, H., More III, B., O'riordan, T., and Svedin, U., (2001) 'Sustainability science', *Science*, 292: 641–642.

Kauffman, S., (1993), *The Origins of Order*, Oxford University Press, New York.

King, A., (1995), 'Avoiding ecological surprise: Lessons from long-standing communities', *Academy of Management Review*, 20: 961–985.

Lambin, E. F., Turner II, B. L., Geist, H. J., Agbola, S. B., Angelsen, A., Bruce, J. W., Coomes, O. T., Dirzo, R., Fischer, G., Folke, C., George, P. S., Homewood, K., Imbernon, J., Leemans, R., Li, X., Moran, E. F., Mortimore, M., Ramakrishnan, P. S., Richards, J. F., Skånes, H., Steffen, W., Stone, G. D., Svedin, U., Veldkamp, T. A, Vogel, C, and Xu, J., (2001), 'The Causes of Land-Use and Land-Cover Change: Moving Beyond the Myths', *Global Environmental Change*, 11:261–269.

Lee, K. N., (1993) *Compass and Gyroscope: Integrating science and politics for the environment*, Island Press, Washington DC.

Levin, S., (1999), *Fragile Dominion: complexity and the commons*, Reading, MA: Perseus Books.

Levin, S., Barrett, S., Aniyar, S., Baumol, W., Bliss, C., Bolin, B., Dasgupta, P., Ehrlich, P., Folke, C., Gren, I.-M., Holling, C. S., Jansson, A.M., Jansson, B.-O., Martin, D., Mäler, K.-G., Perrings, C., and Sheshinsky, E., (1998), 'Resilience in natural and socio-economic systems', *Environment and Development Economics*, 3: 222–235.

Low, B., Ostrom, E., Simon, C., and Wilson, J., (2003) 'Redundancy and diversity in governing and managing natural resources', in F. Berkes, J. Colding and C. Folke (eds), *Navigating Social-Ecological Systems: Building Resilience for Complexity and Change*, Cambridge University Press, Cambridge, UK.

Ludwig, D., Mangel, M., and Haddad, B., (2001), 'Ecology, conservation, and public policy', *Annual Review of Ecology and Systematics*, 32: 481–517.

McGinnis, M., (2000), *Polycentric Governance and Development*, University of Michigan Press, Ann Arbor.

McIntosh, R. J., (2000) 'Social memory in Mande', in R. J. McIntosh, J. A. Tainter and S. K. McIntosh (eds), *The Way the Wind Blows: Climate, History, and Human Action*, pp. 141–180, Columbia University Press, New York.

McIntosh, R. J., Tainter, J. A., and McIntosh, S. K., (eds) (2000), *The Way the Wind Blows: Climate, History and Human Action*, Columbia University Press, New York.

McNeill, J., (2000), *Something New under the Sun: An environmental history of the twentieth century*, The Penguin Press, London.

National Research Council (1998), *Sustaining marine fisheries*, Washington DC: National Academy Press.

Norgaard, R. B., (1994), *Development Betrayed: The End of Progress and a Coevolutionary Revisioning of the Future*, Routledge, New York.

North, D. C., (1990), *Institutions, Institutional Change and Economic Performance*, Cambridge University Press, Cambridge, UK.

Nyström, M., and Folke, C., (2001), 'Spatial resilience of coral reefs', *Ecosystems*, 4: 406–417.

Nyström, M., Folke, C., and Moberg, F., (2000), 'Coral-reef disturbance and resilience in a human-dominated environment', *Trends in Ecology and Evolution*, 15: 413–417.

Olsson, P., and Folke, C., (2001), 'Local Ecological Knowledge and Institutional Dynamics for Ecosystem Management: A Study of Lake Racken Watershed, Sweden', *Ecosystems*, 4: 85–104.

Ostrom, E., (1990), *Governing the Commons: The Evolution of Institutions for Collective Action*, Cambridge University Press, Cambridge.

Ostrom, E., Burger, J., Field, C. B., Norgaard, R. B., and Policansky, D., (1999), 'Sustainability – revisiting the commons: local lessons, global challenges', *Science*, 284:278–82.

Paine, R. T., Tegner, M. J., and Johnson, E. A., (1998), 'Compounded perturbations yield ecological surprises', *Ecosystems*, 1:535–545.

Pearce, D., Markandya, A., and Barbier, E. B., (1989), *Blueprint for a Green Economy*, Earthscan, London.

Petschel-Held, G., Block, A., Cassel-Gintz, M., Kropp, J., Lüdecke, M. K. B., Moldenhauer, O., Reusswig, F., and Schellenhuber, H.-J., (1999) 'Syndromes of global change – a qualitative modelling approach to assist global environmental management', *Environmental Modeling and Assessment*, 4: 295–314.

Pritchard Jr., L., Folke, C., and Gunderson, L. H., (2000), 'Valuation of ecosystem services in institutional context', *Ecosystems*, 3:36–40.

Redman, C. L., (1999), *Human Impact on Ancient Environments*, The University of Arizona Press, Tucson, AZ.

Regier, H. A., and Baskerville, G. L., (1986), 'Sustainable redevelopment of regional ecosystems degraded by exploitive development', in W. C. Clark, R. E. Munn (eds) *Sustainable Development of the Biosphere*, International Institute for Applied Systems Analysis, (IIASA), Cambridge University Press, Cambridge, UK.

Ruddle, K., Hviding, E., and Johannes, R. E., (1992), 'Marine resources management in the context of customary tenure', *Marine Resource Economics*, 7: 249–273.

Scheffer, M., Carpenter, S. R., Foley, J., Folke, C., and Walker, B., (2001), 'Catastrophic shifts in ecosystems', *Nature*, 413:591–696.

Scoones, I., (1999), 'New ecology and the social sciences: what prospects for a fruitful engagement?', *Annual Review of Anthropology*, 28: 479–507.

Shannon, M. A., and Antypas, A. R., (1997), 'Open institutions: Uncertainty and ambiguity in 21st-century forestry', in K. A. Kohm and J. F. Franklin (eds) *Creating a Forestry for the 21st Century: The Science of Ecosystem Management*, Island Press, Washington DC.

Thompson, M., Ellis, R., and Wildavsky, A., (1990), *Cultural Theory*, Westview, Boulder, CO.

Turner, B. L., Clark, W. C., and Kates, R. W., (eds) (1990), *The Earth as Transformed by Human Action: Global and regional changes in the biosphere over the past 300 years*, Cambridge University Press, Cambridge, UK.

van der Leeuw, S., (2000), 'Land degradation as a socionatural process', in R. J. McIntosh, J. A. Tainter and S. K. McIntosh (eds), *The Way the Wind Blows: Climate, History, and Human Action*, Columbia University Press, New York.

Westley, F., (2002), 'The devil in the dynamics: Adaptive management on the front lines', in L. H. Gunderson and C. S. Holling (eds), *Panarchy: Understanding Transformations in Systems of Humans and Nature*, Island Press, Washington, DC.

Index

Printed and bound by CPI Group (UK) Ltd, Croydon, CR0 4YY

21/10/2024

01777083-0001